CELL BIOLOGY RESEARCH PROGRESS

GOLGI APPARATUS: STRUCTURE, FUNCTIONS AND MECHANISMS

CELL BIOLOGY RESEARCH PROGRESS

Additional books in this series can be found on Nova's website under the Series tab.

Additional E-books in this series can be found on Nova's website under the E-books tab.

CELL BIOLOGY RESEARCH PROGRESS

GOLGI APPARATUS: STRUCTURE, FUNCTIONS AND MECHANISMS

CHRISTOPHER J. HAWKINS
EDITOR

Nova Science Publishers, Inc.
New York

Copyright © 2011 by Nova Science Publishers, Inc.

All rights reserved. No part of this book may be reproduced, stored in a retrieval system or transmitted in any form or by any means: electronic, electrostatic, magnetic, tape, mechanical photocopying, recording or otherwise without the written permission of the Publisher.

For permission to use material from this book please contact us:
Telephone 631-231-7269; Fax 631-231-8175
Web Site: http://www.novapublishers.com

NOTICE TO THE READER

The Publisher has taken reasonable care in the preparation of this book, but makes no expressed or implied warranty of any kind and assumes no responsibility for any errors or omissions. No liability is assumed for incidental or consequential damages in connection with or arising out of information contained in this book. The Publisher shall not be liable for any special, consequential, or exemplary damages resulting, in whole or in part, from the readers' use of, or reliance upon, this material. Any parts of this book based on government reports are so indicated and copyright is claimed for those parts to the extent applicable to compilations of such works.

Independent verification should be sought for any data, advice or recommendations contained in this book. In addition, no responsibility is assumed by the publisher for any injury and/or damage to persons or property arising from any methods, products, instructions, ideas or otherwise contained in this publication.

This publication is designed to provide accurate and authoritative information with regard to the subject matter covered herein. It is sold with the clear understanding that the Publisher is not engaged in rendering legal or any other professional services. If legal or any other expert assistance is required, the services of a competent person should be sought. FROM A DECLARATION OF PARTICIPANTS JOINTLY ADOPTED BY A COMMITTEE OF THE AMERICAN BAR ASSOCIATION AND A COMMITTEE OF PUBLISHERS.

Additional color graphics may be available in the e-book version of this book.

LIBRARY OF CONGRESS CATALOGING-IN-PUBLICATION DATA
Golgi apparatus : structure, functions and mechanisms / editor, Christopher J. Hawkins.
 p. cm.
Includes index.
ISBN 978-1-61122-051-3 (softcover)
1. Golgi apparatus. I. Hawkins, Christopher J.
QH603.G6G656 2010
571.6'56--dc22
 2010037903

Published by Nova Science Publishers, Inc. ✦ New York

CONTENTS

Preface		vii
Chapter 1	Golgi Organization and Stress Sensing *Suchismita Chandran and Carolyn Machamer*	1
Chapter 2	Functional Relationships Between Golgi Dynamics and Lipid Metabolism and Transport *Asako Goto and Neale Ridgway*	43
Chapter 3	Signaling Pathways Controlling Mitotic Golgi Breakdown in Mammalian Cells *Inmaculada López-Sánchez and Pedro A. Lazo*	91
Chapter 4	Golgi Apparatus and Hypericin-Mediated Photodynamic Action *Chuanshan Xu, Xinshu Xia, Hongwei Zhang and Albert Wingnang Leung*	105
Chapter 5	Role of the Trans-Golgi Network (TGN) in the Sorting of Nonenzymic Lysosomal Proteins *Maryssa Canuel, Libin Yuan and Carlos R. Morales*	117
Chapter 6	Golgi Apparatus Functions in Manganese Homeostasis and Detoxification *Richard Ortega and Asunción Carmona*	151
Index		161

PREFACE

The Golgi apparatus is an organelle found in most eukaryotic cells. The primary function of the Golgi apparatus is to process and package macromolecules, such as proteins and lipids, after their synthesis and before they make their way to their destination. This book presents topical research data in the study of Golgi apparatus, including Golgi organization and stress sensing; signaling pathways controlling mitotic Golgi breakdown in mammalian cells; the role of Golgi apparatus in the biological mechanisms of hypericin-mediated photodynamic therapy; the role of the Trans-Golgi Network (TGN) in the sorting of nonenzymic lysosomal proteins; and the mechanisms involving the role of Golgi apparatus alteration in neurological disorders triggered by manganese.

Chapter 1 - The eukaryotic Golgi complex plays a central role in processing and sorting of cargo in the secretory pathway. The mammalian Golgi apparatus is composed of multiple stacks of cisternal membranes that are organized laterally into a ribbon-like structure at a juxtanuclear location. The stacks are polarized and protein cargo moves through the organelle in a cis-to-trans direction. In addition, trans-Golgi membranes come in close apposition with specialized endoplasmic reticulum (ER) membranes. These contacts are believed to mediate lipid transfer from the ER directly to the trans-Golgi. The Golgi ribbon structure is unique to vertebrate cells as lower eukaryotic cells lack this elaborate architecture. The complexity of the Golgi ribbon is intriguing and suggests potential additional functions. In this chapter, we discuss the structure of the mammalian Golgi ribbon and its potential role as a sensor of cellular stress. We focus on the role of Golgi organization in ceramide trafficking. Ceramide is a potent secondary messenger in signaling and apoptosis, and its levels are tightly regulated in cells. A protein called CERT (ceramide transfer protein) delivers ceramide from its site of synthesis in the ER to the trans-Golgi for sphingomyelin (SM) synthesis. CERT interacts with both ER and Golgi membranes, and may function at the ER-trans-Golgi contact sites. Some Golgi structural perturbations reduce SM synthesis as well as CERT's colocalization with Golgi markers, suggesting that the organization of the mammalian Golgi ribbon together with CERT may promote specific ER-Golgi interactions for efficient delivery of ceramide for SM synthesis. Under cellular stress, caspase activation can lead to Golgi ribbon disassembly and loss of ER-trans-Golgi contact sites. Prolonged stress that cannot be repaired usually results in apoptosis. Interestingly, increased ceramide levels have been associated with apoptosis, but it is not yet known if newly synthesized ceramide resulting from perturbation of ER-trans-Golgi contact sites contributes to ceramide signaling during apoptosis. An important question is whether ER-trans-Golgi contact sites are upstream targets of stress signals leading to increased ceramide levels and caspase activation, or if altered ceramide trafficking is downstream of Golgi disassembly. Regardless, it is clear that Golgi ribbon structure, including ER-trans-Golgi contact sites is exquisitely sensitive to perturbation, making this organelle an ideal platform to sense cellular stress and integrate signals that determine cell survival or cell death.

Chapter 2 - The Golgi apparatus is a sorting nexus for protein and lipids exported from the endoplasmic reticulum (ER) to other organelles and for secretion. The lipids and sterols that delineate the vesicular/tubular transport carriers and cisternae that constitute the Golgi transport apparatus are just packaging materials but participate directly in membrane fusion, cargo sorting and polarized transport. Low abundance lipids, such as diacylglycerol (DAG), phosphatidic acid (PtdOH), lyso-phospholipids and phosphatidylinositol phosphates, contribute to these processes by localized synthesis and interconversion. These lipids alter the structure of membranes by assisting in induction of positive and negative curvature required for carrier assembly, and regulate the activity of proteins that temporally and spatially regulate fusion and fission events. The Golgi apparatus is especially enriched in phosphatidylinositol 4-phosphate (PtdIns(4)P), where localized metabolism by Golgi-associated PtdIns 4-kinases (PI4K) and phosphatases controls PtdIns(4)P pools that recruit proteins involved in lipid transport and vesicular trafficking. DAG and ceramide conversion in the late Golgi and trans-Golgi network (TGN) by sphingomyelin (SM) synthase regulates Golgi trafficking by recruiting and activating protein kinase D (PKD) for phosphorylation of targets such a PI4KIIIβ. SM and glycosphingolipids (GSL) synthesized in the Golgi apparatus condenses with cholesterol into nanoscale assemblies called lipid rafts. These platforms function in membrane signaling and regulate trans-Golgi network (TGN)-sorting machinery. There is an increasing appreciation for the role of lipid and sterol transfer proteins in modulation of Golgi apparatus function. In particular, site-directed ceramide and sterol transfer proteins that communicate lipid status, and regulate cholesterol, SM and GSL metabolism. Here we will review the highly integrated lipid metabolic and signaling pathways housed in the Golgi apparatus that control secretory activity and membrane assembly.

Chapter 3 - In mitosis, each daughter cell must receive a complete and equal set of cellular components. Cellular organelles that are single copy, such as endoplasmic reticulum, nuclear envelope and Golgi apparatus, have to break down to allow their correct distribution between daughter cells. The mammalian Golgi is a continuous membranous system formed by cistern stacks, tubules and small vesicles that are located in the perinuclear area. At the onset of mitosis, the Golgi apparatus undergoes a sequential fragmentation that is highly coordinated with mitotic progression and in which reversible phosphorylation plays a critical regulatory role. In fact, several kinases have been implicated in each stage of this fragmentation process. Before mitotic disassembly, the lateral connections between the stacks are severed resulting in the formation of isolated cisternae. Several kinases such as mitogen-activated protein kinase kinase 1(MEK1), Raf-1, ERK1c, ERK2, Plk3, VRK1, several Golgi matrix proteins (GRASP65 and GRASP55) and the membrane fission protein BARS have been shown to mediate signals in this first step that takes place in late G2 phase. As prophase progresses, the isolated cisternae are first unstacked followed by its breakage into smaller vesicles and tubules that accumulate around the two spindle poles at metaphase. Unstacking and vesiculation are triggered by several proteins including kinases (Plk1 and Cdc2), the GTPase ARF-1 and inactivation of membrane fusion complexes (VCP and NSF). Post-mitotic Golgi reassembly consists of two processes: membrane fusion mediated by two

ATPases, VCP and NSF; and cistern restacking mediated by dephosphorylation of Golgi matrix proteins (GRASP65 and GM130) by phosphatase PP2A (Bα). Apart from the tight regulation by reversible phosphorylation, it seems that mitotic Golgi membrane dynamics also involves a cycle of ubiquitination during disassembly and deubiquitination during reassembly in part regulated by the VCP-mediated pathway.

Chapter 4 - Hypericin isolated from *Hypericum perforatum* plants, exhibits a wide range of biological activities and medical applications for treating tumors. Emerging evidence has demonstrated that hypericin could be activated by visible light to produce reactive oxygen species (ROS) which destroys a tumor directly or indirectly. Hypercin-induced photodynamic therapy has showed considerable promise as an alternative modality in the management of malignant tumors. However, the exact mechanisms need to be clarified. Recent studies have showed that hypericin was accumulated in the Golgi apparatus, indicating that the role of the Golgi apparatus is indispensable in the biological mechanisms of hypericin-mediated photodynamic therapy.

Chapter 5 - In eukaryotes the delivery of newly synthesized proteins to the extracellular space, the plasma membrane and the endosome/lysosomal system is dependent on a series of functionally distinct compartments, including the endoplasmic reticulum, the Golgi apparatus and carrier vesicles. This system plays a role in the post-translational modification, sorting and distribution of proteins to their final destination. Most cargo is sorted within, and exits from, the *trans*-Golgi network (TGN). Proteins delivered to the endosomal/lysosomal system include a large and diverse class of hydrolytic enzymes and nonenzymic activator proteins. They are directed away from the cell surface by their binding to mannose-6-phosphate receptors (MPR). Surprisingly, in I-cell disease, in which the MPR pathway is disrupted, the nonenzymic sphingolipid activator proteins (SAPs), prosaposin and GM_2AP, continue to traffic to the lysosomes. This observation led us to the discovery of a new lysosomal sorting receptor, sortilin. Both prosaposin and GM_2AP are secreted or targeted to the lysosomes through an interaction of specific domains with sortilin. In the case of prosaposin, deletion of the C-terminus did not interfere with its secretion, but abolished its transport to the lysosomes. Our investigations also showed that while the lysosomal isomer of prosaposin (65kDa) is Endo H-sensitive, the secretory form (70kDa) is Endo H-resistant. Since the processing pathway within the Golgi apparatus is highly ordered, this Endo-H analysis permitted us to distinguish a sorting sub-compartment where a significant fraction of prosaposin exits to the endosomal/lysosomal system prior to achieving full glycosylation and Endo-H resistance. Mutational analysis revealed that the first half of the prosaposin C-terminus (aa524-540) contains a saposin-like motif required for its binding to sortilin and its transport to the lysosomes. Additionally, a chimeric construct consisting of albumin and a distal segment of prosaposin, which included its C-terminus, resulted in the routing of albumin to the lysosomes. Based on previous observations showing that the lysosomal trafficking of prosaposin and chimeric albumin required sphingomyelin, we tested the hypothesis that these proteins, as well as sortilin, are associated with detergent-resistant membranes (DRMs). Our results demonstrated that indeed sortilin, prosaposin and chimeric albumin are found in DRMs, and that the sorting of prosaposin to DRMs depends upon the interaction of its C-terminus with sortilin. In conclusion, we have identified a specific segment in the C-terminus of prosaposin, as well as amino acid residues that are critical to the binding of prosaposin to sortilin and its subsequent lysosomal trafficking. The identified sequence may permit the development of new therapeutic approaches for the targeting of proteins with anti-pathogenic properties to penetrate the cell via the endocytic pathway.

Chapter 6 - Recent data suggest that the Golgi apparatus plays a key role in the homeostasis and detoxification of manganese. Manganese is an essential trace element but when high exposure conditions occur, manganese induces neurological symptoms in human. Manganese is also suspected to be an environmental risk factor in the aetiology of Parkinson's disease. However, the mechanisms regulating manganese homeostasis and detoxification in mammalian cells are largely unknown. Owing to the development of synchrotron radiation X-ray nano-chemical imaging, we revealed the specific accumulation of manganese in the Golgi apparatus of dopaminergic cells in culture.

At both physiological and subcytotoxic concentrations of manganese, we found that manganese was essentially located within the Golgi apparatus. At cytotoxic concentration of manganese, we found a large increase of manganese content in the cytoplasm and the nucleus of dopaminergic cells. Similarly, if the Golgi apparatus is altered using brefeldin A, manganese reaches the nucleus and cytoplasm in higher content. The accumulation of manganese in the Golgi apparatus could have a preventative effect because manganese could be removed by exocytosis. However, vesicular trafficking could be disturbed by high concentrations of manganese leading to neuronal cell death. We will discuss the mechanisms involving the role of Golgi apparatus alteration in neurological disorders triggered by manganese.

In: Golgi Apparatus: Structure, Functions…
Editor: Christopher J. Hawkins, pp. 1-41

ISBN 978-1-61122-051-3
© 2011 Nova Science Publishers, Inc.

Chapter 1

GOLGI ORGANIZATION AND STRESS SENSING

Suchismita Chandran and Carolyn Machamer
Department of Cell Biology
Johns Hopkins University School of Medicine, USA

ABSTRACT

The eukaryotic Golgi complex plays a central role in processing and sorting of cargo in the secretory pathway. The mammalian Golgi apparatus is composed of multiple stacks of cisternal membranes that are organized laterally into a ribbon-like structure at a juxtanuclear location. The stacks are polarized and protein cargo moves through the organelle in a cis-to-trans direction. In addition, trans-Golgi membranes come in close apposition with specialized endoplasmic reticulum (ER) membranes. These contacts are believed to mediate lipid transfer from the ER directly to the trans-Golgi. The Golgi ribbon structure is unique to vertebrate cells as lower eukaryotic cells lack this elaborate architecture. The complexity of the Golgi ribbon is intriguing and suggests potential additional functions. In this chapter, we discuss the structure of the mammalian Golgi ribbon and its potential role as a sensor of cellular stress. We focus on the role of Golgi organization in ceramide trafficking. Ceramide is a potent secondary messenger in signaling and apoptosis, and its levels are tightly regulated in cells. A protein called CERT (ceramide transfer protein) delivers ceramide from its site of synthesis in the ER to the trans-Golgi for sphingomyelin (SM) synthesis. CERT interacts with both ER and Golgi membranes, and may function at the ER-trans-Golgi contact sites. Some Golgi structural perturbations reduce SM synthesis as well as CERT's colocalization with Golgi markers, suggesting that the organization of the mammalian Golgi ribbon together with CERT may promote specific ER-Golgi interactions for efficient delivery of ceramide for SM synthesis. Under cellular stress, caspase activation can lead to Golgi ribbon disassembly and loss of ER-trans-Golgi contact sites. Prolonged stress that cannot be repaired usually results in apoptosis. Interestingly, increased ceramide levels have been associated with apoptosis, but it is not yet known if newly synthesized ceramide resulting from perturbation of ER-trans-Golgi contact sites contributes to ceramide signaling during apoptosis. An important question is whether ER-trans-Golgi contact sites are upstream targets of stress signals leading to increased ceramide levels and caspase activation, or if altered ceramide trafficking is downstream of Golgi disassembly. Regardless, it is clear that Golgi ribbon structure, including ER-trans-Golgi contact sites is exquisitely sensitive to perturbation, making this organelle an ideal platform to sense cellular stress and integrate signals that determine cell survival or cell death.

1. INTRODUCTION

The Golgi complex is an important processing and sorting organelle in eukaryotic cells. It is present in a juxtanuclear region and forms the central hub of the secretory pathway. Newly synthesized proteins and lipids transported from the endoplasmic reticulum (ER) to the Golgi are post-translationally modified and sorted to be trafficked to different destinations in the cell, including the plasma membrane, endosomes and lysosomes. In addition to supporting vesicular transport, resident Golgi enzymes modify cargo by glycosylation, phosphorylation, sulfation and proteolytic processing. The Golgi also plays an important role in synthesis of glycosaminoglycans that link with core proteins to form proteoglycans. Importantly, the Golgi is involved in synthesis of phospholipids and sphingolipids, including sphingomyelin, glucosylceramide and gangliosides. In addition, the Golgi complex supports endocytosis and membrane recycling events. The Golgi apparatus also supports various cellular functions, including cell migration, cell cycle and apoptosis.

Some functions of the Golgi are closely linked to its structure. The fundamental structure of the Golgi apparatus consists of flattened cisternal membranes that are associated with numerous vesicles. The cisternae are organized into stacks, and in mammalian cells the stacks are connected laterally to form a continuous membrane system called the Golgi ribbon that is held in a perinuclear region by microtubules. The stacks are polarized with the side receiving cargo from the ER called cis-Golgi and the side releasing cargo called trans-Golgi. The stacks are functionally compartmentalized and include the cis-Golgi network (CGN), the cis-, medial-, trans-cisternae and the trans-Golgi network (TGN) [1]. The CGN, also called the ER-intermediate compartment (ERGIC) and the TGN were originally defined by the exaggeration of their respective membranes upon temperature block during membrane trafficking. At temperatures below $16^{\circ}C$, cargo accumulates in the CGN compartment, while at $20^{\circ}C$, cargo accumulates in the TGN [2,3,4,5,6,7,8]. The CGN and the TGN participate in sorting lipids and proteins during membrane trafficking [9,10,11]. In addition, the TGN has been implicated in membrane recycling events and endocytosis. Interestingly, some morphological and functional features of the CGN are similar to the endocytic recycling compartment, a centrally located subpopulation of the endocytic pathway. Thus, the CGN can be termed the biosynthetic recycling compartment, a long-lived mirror image of the endocytic recycling compartment [12]. Both compartments function in membrane recycling in response to cellular processes including migration and differentiation. The enzymes that are involved in processing cargo are localized within Golgi subcompartments in order of their mode of action. For example, enzymes that are involved in processing glycoproteins are distributed in each cisternal subcompartment according to the order of modification of sugar chains, with enzymes that are involved in removal of mannose sugar residues residing in cis-/medial-Golgi subcompartments, while enzymes that add N-acetylglucosamine and galactose residing in medial- and trans-Golgi compartments, respectively [13,14]. Thus the cisternal subcompartments mediate partitioning of different Golgi resident enzymes.

The Golgi apparatus receives protein and lipid cargo from the ER and traffics these molecules through the different subcompartments in a cis-to-trans fashion [15,16], and reviewed in [17]). There are two opposing views on how the Golgi participates in the secretory pathway: the vesicular transport model predicts that the Golgi is a stable structure where cargo is transported through the Golgi cisternae by vesicles that bud from one membrane and fuse with another [16], while the cisternal maturation model predicts that the Golgi stack is a transitional structure. In the latter model, cargo emerging from the ER remains in cisternae, while each cisterna matures as resident enzymes are transported backwards by vesicles [18,19]. Since there is evidence that both intra-Golgi vesicular transport and cisternal maturation occur in mammalian cells (depending on cell type and cargo), it is likely that a combination of these mechanisms may be used to traffic cargo through the Golgi [20,21].

As previously described, the Golgi ribbon structure is unique to vertebrate cells as lower eukaryotic systems lack this level of organization. In the budding yeast *Saccharomyces cerevisiae*, the Golgi is composed of multiple cisternae that are dispersed throughout the cell. The cisternae are not arranged into stacks, but appear to be polarized, with subsets of cisternae possessing specific Golgi resident enzymes. However, other yeasts (including *Pichia pastoris* and *Schizosaccharomyces pombe*) do have stacked cisternal Golgi membranes. The reason why the Golgi is organized into stacks in some species of yeast but not others remains unclear. However, molecular evolutionary evidence suggests that the ancestral eukaryote possessed a stacked Golgi and that some of the organisms underwent a Golgi unstacking process during evolution [22]. In most *Drosophila* cells, the Golgi cisternae are organized into polarized stacks, but the stacks are not laterally connected. Instead they are scattered throughout the cytoplasm. However in the onion stage spermatids of *Drosophila*, a Golgi ribbon called the acroblast is seen in a perinuclear region (reviewed in [23]). Thus, the machinery to build the Golgi ribbon in *Drosophila* exists, but the reason why *Drosophila* cells predominantly lack Golgi ribbon organization or why the Golgi ribbon appears only in certain cells during certain developmental stages is not known [23]. Interestingly, some developmental stages of *Drosophila* including early embryogenesis and imaginal discs from early/mid third instar lava do not even exhibit Golgi stacking. Instead, the Golgi apparatus consist of clusters of vesicles and tubules (reviewed in [23]).

It is thought that the mammalian Golgi ribbon supports efficient processing and sorting of cargo. However, the secretory pathway is robust in *Drosophila* and yeast cells where not only a Golgi ribbon is lacking but also where Golgi stacks are absent [23,24]. Perhaps the existence of the Golgi ribbon in mammalian cells points towards additional functions that have not yet been identified.

In this chapter, we discuss the structure of mammalian Golgi ribbon and compare its complex organization with that of lower eukaryotes. We speculate that the complex nature of the mammalian Golgi points to additional functions that have not been identified or characterized. We also focus on the concept of the mammalian Golgi as a potential sensor of cellular stress, and as a platform for transducing downstream signaling in the events that lead to cell survival or cell death, particularly with regard to ceramide trafficking.

2. Golgi Structure

The mammalian Golgi apparatus is centrally positioned and is maintained by its association with the microtubule organizing center (MTOC) or the centrosome, the actin cytoskeleton and the ER. In order to understand the unique functions of the mammalian Golgi apparatus it is important to understand its unique architectural design and subcellular location. In this section we review the structural components of the Golgi apparatus and its complex organization.

2.a. The Intricacies of the Mammalian Golgi Apparatus

Much of our understanding of Golgi structure and function comes from biochemical and molecular studies, immunofluorescence and conventional electron microscopy (EM). Three-dimensional (3-D) extrapolation from two-dimensional images of Golgi structure [25,26,27] could not provide enough information to advance the Golgi structure-function model. However, development of dual axis, high voltage EM (HVEM) tomography, an advanced tool to visualize cellular morphology in 3-D using EM, has allowed additional insights into the structural organization of the Golgi complex [27]. The information from HVEM tomography in mammalian cells (including normal rat kidney cells, insulin secreting pancreatic beta cell lines, and mammary and pancreatic tissue), not only confirmed previous observations of Golgi structure but also revealed structural details of the Golgi ribbon that implicate additional functions [27,28,29,30,31]. In this segment we highlight some of the findings on Golgi ribbon structure as revealed by HVEM (summarized in Figure 1).

In mammalian cells, the Golgi ribbon is comprised of many stacks, with each stack made up of multiple cisternae that are fenestrated and associated with coated buds [28,32]. The CGN, which receives cargo from the ER, is made up of distinct, discontinuous polymorphic membranous structures and is situated in front of the cis-most Golgi cisterna [28]. The CGN is also characterized by the presence of structures similar to budding and fusing vesicles that distinctly lack a coat. Some of the CGN elements are shown to exist as discrete entities, while some of them appear to fuse and dock at the cis-most cisternae [27,28]. The cisterna at the cis-most side of the Golgi that receives cargo from the CGN is highly fenestrated but continuous along the Golgi ribbon [27,28].

The fenestrations in the cis-most cisterna were found to align with most of the fenestrations of the adjacent cisterna [27,28]. The trans-side of the Golgi is characterized by the presence of clathrin-coated vesicles, which suggests that exit of cargo from the Golgi occurs from the trans-Golgi [29]. The TGN is situated trans of the Golgi stack and is thought to be predominantly a tubular compartment [29]. It has been depicted as continuous with either only the last trans-Golgi cisterna [29,33] or multiple trans-cisternae of the Golgi stack [26,29]. However, Ladinsky et al show by HVEM tomography that the Golgi apparatus lacks a distinct TGN [29]. Instead highly fenestrated tubules extend from the ultimate and penultimate trans–cisternae, suggesting that sorting and exit from the Golgi can occur from multiple trans-cisternae. In addition, outward extending tubules with budding tips were shown to emerge from the margins of both the cis- and trans-Golgi cisternae, not just the trans-cisternae, and appear to be perpendicular to the plane of the cisternae [28].

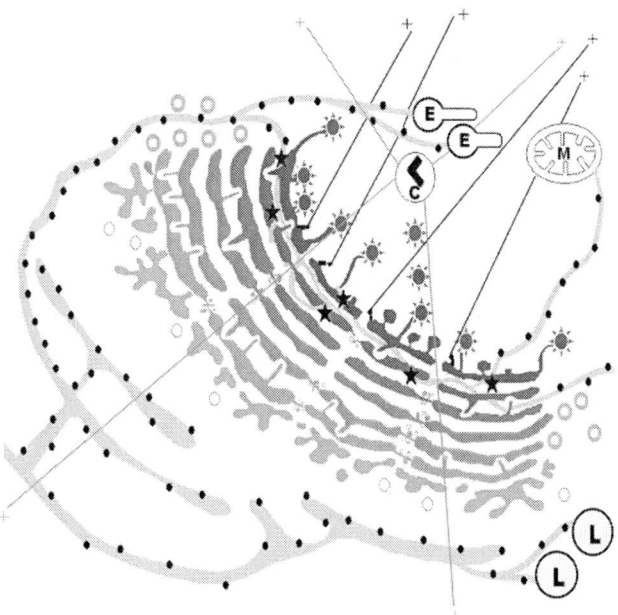

Figure 1. Characteristics of the Golgi ribbon as viewed by HVEM. The Golgi ribbon consists of multiple cisternae (shown in various shades of grey) that are organized into stacks (4 stacks are shown in the cartoon). The stacks are arranged adjacent to each other and equivalent cisternae are laterally connected to form a continuous membrane. The lateral connections are characterized by a non-compact region that contains tubules and polymorphic membranes. The CGN on the cis side (lightest shade of grey in the Golgi stack) is characterized by discontinuous polymorphic structures composed of non-coated vesicles and tubules that either exist as discrete entities or can dock with the adjacent cis-Golgi membrane. The cis-most cisterna of the Golgi is highly fenestrated. COP-coated vesicles (circles with grey border) can bud and fuse from the cis- and medial-Golgi cisternae. The trans-most side of the Golgi (darkest shade of grey in the Golgi stack) is tubulated and distinguished by clathrin-coated vesicles (circles with spikes). ER membranes (studded with ribosomes, represented by small filled circles) come to lie in close apposition with trans-Golgi membranes to form trans-ER-trans-Golgi membrane contact sites that may be stabilized by lipid transfer proteins including CERT and OSBP (indicated by stars). The ER membrane also comes in close apposition with other cellular organelles like endosomes (E), lysosomes (L), and mitochondria (M). CLASP-nucleated microtubules that arise from the Golgi are indicated by black solid lines, while microtubules that arise from the centrosome (C) are indicated by grey solid lines. The plus ends of microtubules are indicated (+).

Three-D reconstruction studies with HVEM also show that the stacks that make up the mammalian Golgi apparatus are laterally connected, giving the organelle its characteristic ribbon-like appearance [28]. The stacks are separated by a non-compact region that is composed of vesicles and polymorphic membranous structures that make connections between equivalent cisternae from different stacks [28]. Thus, the Golgi apparatus shows a dual level of complex organization: organization of cisternae into stacks, and organization of stacks into a ribbon.

The Golgi ribbon as described above is characteristic of the interphase stage of the cell cycle, and is held in a juxtanuclear region by microtubules that emanate from the adjacent centrosome. But as the cell progresses into mitosis, the Golgi ribbon disassembles and the fragmented Golgi disperses throughout the cell [34,35]. The disassembly of the mitotic Golgi is mediated by depolymerization of microtubules as well as phosphorylation of several Golgi-associated proteins that maintain its structure including GRASP55, GRASP65 and GM130, which are described below [35,36,37]. During this time, membrane trafficking is blocked due to continuous budding of transport vesicles and an inhibition of vesicle fusion with target membranes [37]. However, during telophase and cytokinesis, the Golgi cisternal membranes interconnect and the Golgi ribbon structure is inherited by the daughter cells [38]. The re-establishment of the Golgi ribbon in daughter cells is once again dependent on microtubules and Golgi structural proteins. Interestingly, inhibition of Golgi disassembly during mitosis leads to cell cycle arrest in the G2 phase of the cell cycle, whereas induction of Golgi disassembly during G2 is sufficient to overcome the arrest, suggesting the existence of a Golgi checkpoint during cell cycle events [36,39,40,41]. CtBP1-S/BARS (C-terminal binding protein 3/Brefeldin A adenosine diphosphate-ribosylated substrate), also called BARS, is a protein involved in severing the non compact region and lateral links of the Golgi ribbon and is required for mitotic entry [40]. The Golgi ribbon is disassembled into isolated stacks during G2 in a BARS-dependent manner before cells proceed through the G2/M checkpoint [40]. Cells lacking a Golgi ribbon enter mitosis independently of BARS [40]. Embryonic fibroblasts from BARS knockout mice show lack of a Golgi ribbon at all stages of cell cycle, and do not require BARS for mitotic entry. [40]. Taken together, these studies suggest that the Golgi ribbon may be involved in functions other than membrane trafficking, specifically in cell regulatory events. Thus, the organization of the Golgi ribbon is a dynamic process and understanding its assembly and disassembly process may help unravel some of the unidentified functions of the Golgi apparatus.

2.b. Golgi Ribbon Structure and Interaction with Other Organelles

To further complicate the architectural organization of the Golgi apparatus, the organelle is found to be closely associated with ER membranes. The ER not only makes contact at the cis-side of the Golgi to form the CGN or ERGIC [9,10,11,27,28], but also makes contact with the trans-side of the Golgi (Figure 1). Careful examination by HVEM tomography has revealed a close association between ER and trans-Golgi membranes, called membrane contact sites [27,28]. This intimate ER-Golgi relationship was previously identified as the GERL (Golgi-ER- lysosome) by Alex Novikoff and was implicated in lysosome production [42,43]. The GERL theory was later discarded, but the close apposition between ER and trans-Golgi membranes was confirmed by HVEM, and the contact sites are now designated as ER-trans-Golgi membrane contact sites. Three-D reconstruction studies revealed that the ER traverses the Golgi cisternae and is intimately wrapped around swollen regions of the organelle [27]. Some regions of the traversing ER contain bound ribosomes while other regions lack ribosomes [30]. Interestingly, at several points, the ER membrane was found to traverse the entire Golgi stack through openings in the cisternae and extend beyond the cis- and trans-cisternae [30]. The openings in the Golgi cisternae through which ER tubules pass are also thought to be the points through which tubules from cisternae traverse the Golgi stack, thus connecting the cisternae within each stack [30]. However, the cisternal tubules and the ER membranes that pass through these openings are thought to remain distinct from one another [30]. This intimate relationship between the ER and the trans-Golgi is thought to mediate trafficking of lipids from the ER directly to the trans-Golgi, bypassing the vesicular route thorough which the bulk flow of cargo is mediated. It is also noteworthy that the ER membranes that make close contact with Golgi membranes are continuous and are also intimately associated with mitochondria, endosomes and lysosomes [30], perhaps suggesting a less characterized but direct cargo trafficking route between all the organelles involved. This finding indicates that the Golgi apparatus does not function as a discrete organelle in cells, but that its structure, dynamics and function are all tied into the structure and function of other cellular organelles. Thus, it is very likely that the ER-trans-Golgi contact sites contribute towards maintenance of Golgi ribbon structure and any perturbation to these contact sites could result in disruption to Golgi structure and function. Furthermore, it is also possible that perturbation of Golgi structure, including disassembly of Golgi stacks by perturbing Golgi structural proteins or altering membrane lipid dynamics could result in disruption of ER-Golgi contact sites and their role in non-vesicular trafficking. Since a continuous ER membrane making contact sites with trans-Golgi cisternae, mitochondria, endosomes, and lysosomes was seen [30], it can be speculated that perturbation of the structure or function of other subcellular organelles may also affect Golgi structure and function. This could occur either via disruption in the ER structure or by altered signaling or block in non-vesicular cargo trafficking. The later possibility is intriguing, but currently very little data exist in this area as no structural components have been discovered that define membrane contact sites. Hence, it is difficult to predict how other subcellular organelles might contribute towards maintenance and regulation of the various aspects of the Golgi apparatus via the ER contact sites. Interestingly, no membrane contact sites have been identified in *Saccharomyces cerevisiae* [44]. In *Drosophila* cells, the dispersed Golgi stacks are thought to make contact with the ER at their cis-side and are thought to be closely associated with ER exit sites [23,45]. However, it is unclear if trans-Golgi membrane from the dispersed *Drosophila* Golgi stacks associate with the ER to form trans-ER-trans-Golgi membrane contact sites.

2.c. Golgi Structure and Stacking

The mechanisms that mediate the complicated architecture of the Golgi ribbon are not very well understood. Golgins and Golgi reassembly stacking proteins (GRASPs) have been implicated in Golgi stacking and ribbon maintenance [46,47]. GRASP-65, a cis-Golgi localized [48] peripheral Golgi protein was identified as the first component required for Golgi membrane stacking in an in vitro assay for Golgi disassembly and reassembly [49]. Subsequently, the role of GRASP-65 in Golgi stacking was confirmed in mammalian cells [50,51] and *Drosophila* tissues [52]. Furthermore, GRASP-65 is capable of forming stable homodimers that can interact with homodimers from adjacent cisternal membranes to form oligomers, resulting in cisternal stacking [50,53]. Phosphorylation inhibits oligomerization of GRASP-65 and can thus contribute towards Golgi disassembly during mitosis [54,55]. A recent study reported that GRASP-65 acts collaboratively with its medial-Golgi [48] homologue GRASP-55 in Golgi stacking [51], suggesting that GRASP oligomerization may play a central role in Golgi structure in interphase mammalian cells. However, a separate study implicated GRASP-65 in linking stacks to form the Golgi ribbon [56]. Regardless of whether GRASPs mediate Golgi stacking and/or ribbon formation, it is clear that this group of proteins is important for higher order organization of Golgi structure. It is interesting to note that although the budding yeast *Saccharomyces cerevisiae* has a GRASP homologue, its

Golgi apparatus is not organized into stacks or ribbon, but instead exists as individual cisternae scattered through the cell. Since yeast GRASP lacks the domain that mediates oligomerization, it is possible that yeast GRASP evolved to perform a different function from mammalian cells [56]. It is also important to mention that although plants have Golgi stacks, they lack GRASP homologues, with likelihood that a highly divergent GRASP-like oligomerizing plant protein may still remain undiscovered [51]. And if GRASP-65 is indeed important for linking Golgi stacks laterally into a ribbon [56], it is intriguing that Drosophila cells lack the ribbon-level of Golgi organization because in most Drosophila tissues, Golgi stacks are scattered throughout the cell. It is possible that the sole Drosophila GRASP protein has evolved to only mediate stacking of Golgi cisternae and not to maintain lateral connectivity between stacks. We now know how GRASPs are regulated during mitosis by phosphorylation [51,54,55], but it would also be interesting to understand how GRASP-65 ties into Golgi ribbon re-organization during muscle differentiation, where the Golgi ribbon is fragmented into stacks scattered throughout the cytoplasm in the multinucleate cells [57]. Thus, understanding the function of GRASPs in various eukaryotic systems might not only be important in understanding the existence of varying Golgi morphologies but may also prove to be key in unraveling the relationship between Golgi structure and function.

The Golgi apparatus maintains a stable architecture during vesicular trafficking events when there is intense membrane flux [58,59]. This stability is contributed by a group of Golgi associated proteins called golgins. Members of the golgin family were originally identified as autoantigens in a wide range of autoimmune diseases, but now include any Golgi-localized protein that contains a long coiled-coiled domain [60]. The golgin family includes peripheral membrane proteins such as GM130 [61], golgin-160 [62,63], golgin-97 [64], golgin-210 [65,66], golgin-230/245 [63,67], GCC (Golgi-localized coiled-coiled protein)-88 [68] and GCC-185 [68] as well as integral membrane proteins such as golgin-45 [69], golgin-67 [70], golgin-84 [71] and giantin [72]. Some golgins participate in vesicle tethering and Golgi structure maintenance. The cis-acting golgin-210 has been implicated in asymmetric tethering of highly curved liposomes to flatter ones, suggesting a role in Golgi architecture maintenance during vesicular trafficking events [73]. The lipid-binding amphipathic lipid-packing sensor at the N-terminus of this protein along with a guanosine triphosphate activating protein for Arf1 (ArfGAP1) is thought to sense membrane curvature during budding and fusion events at the Golgi [73]. Giantin and GM130, two of the most extensively studied golgins, are thought to interact with each other at the cis-side of the Golgi via p115 in a Rab-1 dependent manner, and are thought to promote SNARE (soluble NSF-attachment proteins receptor) mediated vesicle fusion events [74,75]. A number of other golgins have been implicated in maintenance of Golgi structure. Depletion of the medial-Golgi protein golgin-45 by RNA interference results in disassembly of Golgi structure and in inhibition of the secretory pathway [69]. Similarly, depletion of golgin-97 by RNA interference leads to Golgi structural perturbation and block in endosome-to-TGN retrograde trafficking of the cholera toxin B subunit [76]. Also, depletion of golgin-84 results in Golgi fragmentation and partial inhibition of forward trafficking of vesicular stomatitis virus G protein to the plasma membrane [77]. However, the contribution of golgin-160 in Golgi ribbon maintainence is controversial. Although one study reported that depletion of golgin-160 leads to collapse of the Golgi ribbon and dispersal of stacks [78], another set of studies did not observe this phenotype [79,80]. Since golgin-160 is alternatively spliced [81], Yadav et al. [78] suggested that different isoforms of golgin-160 were targeted by RNA interference in the two separate studies, and that the two isoforms could play different roles at the Golgi. Taken together, these studies suggest that golgins and their interacting proteins are not only important for mediating various trafficking events, but are also directly involved in maintaining Golgi structure. Hence, the idea emerges that Golgi structure is intimately involved with its function, and that perturbing one would affect the other. It is interesting to note that while all the yeast golgins thus identified have human orthologues, not all human golgins have a corresponding protein in yeast [82]. Since in yeast the Golgi is organized into dispersed cisternae, it is likely that the presence of additional mammalian golgins may serve to order the Golgi into higher order stacks and ribbon, and also cater to additional cellular functions not present in yeast, perhaps in stress sensing and apoptosis.

2.d. The Cytoskeleton and Maintenance of Golgi Ribbon Structure

The elaborate architecture of the Golgi ribbon is not only maintained by an array of Golgi localized structural proteins, but also by the associated cytoskeletal elements. Both microtubules and the actin cytoskeleton are known to play active roles within the secretory system during membrane trafficking [83,84,85,86]. However, their roles in maintaining and organizing the Golgi structure are not completely understood. Three-D reconstructions of the mammalian Golgi by HVEM tomography reveals that some microtubules are in close association with Golgi cisternae, but the association was particularly enriched at the cis-most region of the Golgi, where the association extended over a considerable distance [32]. In addition, microtubules were found to traverse Golgi stacks through the non-compact regions and cisternal fenestrations at several points [32], thereby pointing towards additional structural details of the Golgi complex.

2.d.1. Microtubules Maintain Golgi Ribbon Structure

The Golgi-ribbon in mammalian cells is closely associated with the microtubule organizing center (MTOC) and is held in a juxtanuclear region by microtubules. Microtubule disassembly causes the Golgi to disassemble into fragments called ministacks [87,88]. The ministacks maintain their polarity and adapt after a short time to function properly to glycosylate and traffic cargo [87,89]. In eukaryotic cells, centrosomes serve as the chief MTOC, from which microtubules emanate. The minus ends of the microtubules are embedded within the centrosome and the dynamic plus ends are extended towards the plasma membrane. Cargo vesicles are transported from ER exit sites to the Golgi apparatus along microtubules towards their minus ends by the molecular motor dynein [90,91], while traffic out of the Golgi complex along microtubules is mediated by predominantly the plus-ended kinesin motor proteins [92,93,94]. Several microtubule-interacting proteins on the cytoplasmic face of cis-Golgi membranes help anchor the Golgi in its juxtanuclear region [65,95]. Interestingly, recent studies have shown that the Golgi apparatus can itself serve as a platform for organizing microtubules and functions as an MTOC, suggesting that microtubule nucleation can occur in a centrosome independent manner [96,97]. These studies demonstrated that in addition to γ-tubulin, microtubule nucleation at the Golgi requires the

TGN localized CLASP proteins. The Golgi-emanating microtubules could contribute towards cell polarization, directional cell migration [98,99,100], and in baso-apical vesicular transport [101,102]. Supporting this, Miller et al. have shown that Golgi nucleated microtubules are important for assembling Golgi stacks into a continuous ribbon [99].

In addition, a separate study reported that the association of microtubules with the Golgi is important for maintaining Golgi structure during mitosis [103]. During mitosis it has been observed that there is partial accumulation of Golgi membranes at the spindle poles [104], while the rest of the membranes are dispersed throughout the cytoplasm [105]. In support of this, when Wei et al. induced asymmetric division in cells such that the entire spindle segregated into only one daughter cell, they saw formation of a Golgi ribbon [103]. However, in daughter cells devoid of spindles, scattered Golgi stacks that were still polarized and capable of cargo transport persisted. Golgi ribbon formation was rescued upon microinjection of a Golgi extract along with tubulin or spindle material, suggesting that Golgi stacks are partitioned by an independent, yet unidentified mechanism, while Golgi ribbon inheritance is contributed by microtubules. It remains to be seen which population of microtubules mediate Golgi ribbon inheritance: the microtubules that are nucleated and maintained at the Golgi or the microtubules that emanate from the centrosomal MTOC, or both. Regardless, these studies indicate that the Golgi structural organization is closely associated and is dependent on the microtubule cytoskeleton for maintaining its structure and in its juxtanuclear positioning. In the yeast *S. cerevisiae* where the Golgi exists as dispersed cisternae, there is a lack of microtubule-dependent organization. Similarly most *Drosophila* tissues also lack this dependence on microtubules, since their Golgi is organized as dispersed stacks. This suggests that the juxtanuclear positioning of the Golgi apparatus in mammalian cells is not essential for processing and sorting of cargo, but could play a role in additional cellular events not yet identified, perhaps in signaling events including stress sensing and apoptosis.

2.d.2. *The Actin Cytoskeleton and Golgi Structure*

The actin cytoskeleton also plays an important role in Golgi structure and function. There is evidence that a population of short actin filaments associates with the Golgi apparatus and Golgi-derived transport vesicles [106]. These short actin microfilaments are characterized by Tm5NM-2, an isoform of tropomyosin [106,107]. In addition, actin filaments along with myosin motors are known to participate in vesicular trafficking from the Golgi, particularly in polarized cells [85]. Although microtubules are required for efficient transport of vesicles to the apical surface of polarized cells, microtubules do not extend into the apical surface, and terminate in a region just below the actin-rich apical domain [108]. Instead, transport vesicles are trafficked along actin filaments by myosin I motor protein in this region of the cell [109,110,111]. The recruitment of actin machinery, including transient recruitment of nonmuscle myosin II to the TGN during budding of transport vesicles has implicated the actin cytoskeleton in mediating vesicular trafficking from the Golgi as well as in maintaining Golgi structure [85,109]. Disruption of microfilaments results in Golgi disassembly where the stacks are still associated with microtubules [112,113].

Many actin binding proteins including tropomyosin [106], centractin [114] as well as several isoforms of β-spectrin [115,116,117] are localized to the Golgi apparatus, suggesting a role in maintaining Golgi structure and function. Spectrin is thought to regulate membrane organization, stability and shape by simultaneously interacting with specific phospholipids, Golgi integral membrane proteins and cytosolic proteins to form a 3-D lattice at the cytoplasmic face of the Golgi apparatus [118]. Depletion of spectrin causes disassembly of the Golgi ribbon in some cell types, and inhibition of protein trafficking [119]. It is interesting to note that the yeast *S. cerevisiae*, which has dispersed Golgi cisternal membranes, also lacks spectrin. In this regard, it would be interesting to understand how spectrin contributes to higher order Golgi assembly in mammalian cells. A number of other proteins that modulate the actin cytoskeleton also localize to the Golgi region. A recent study suggests that ADF (actin depolymerizing factor)/cofilin trimming of the actin cytoskeleton at the Golgi apparatus generates a sorting domain at the TGN, and that inhibition of actin trimming contributes to cargo missorting [120]. Also, Salvarezza et al. indicate that LIM kinase and cofilin organize and maintain a Golgi pool of actin and is important for regulating protein trafficking at the TGN [121]. Moreover, F-actin depolymerization in mammalian cells resulted in a compact Golgi pattern, albeit without disruption to Golgi stacking or post-Golgi trafficking events [112,113,122]. And interestingly, a separate study reported the involvement of F-actin depolymerization in Golgi segregation during mitosis in *Drosophila* [123]. Here, the authors show that Golgi stacks duplicate to form an actin mediated paired structure during G1/S phase that depends on Abi (Abl interactor)/Scar (suppressor of cAMP receptor) activity. Abi and Scar are two cytoskeletal proteins that modulate actin dynamics by activating Arp2/3 (actin-related protein 2/3)-dependent actin nucleation. During G2 phase, the two stacks separate just before mitosis. Preventing separation of the paired Golgi stacks during G2 phase inhibits entry into mitosis, suggesting that the paired organization of the Golgi apparatus is part of the mitotic checkpoint in *Drosophila*. Since the structural organization of the Golgi in both *S. cerevisiae* and *Drosophila* is not influenced by microtubules but is dependent on the actin cytoskeleton [123,124], it is likely that the actin cytoskeleton plays a fundamental role in Golgi organization.

Although the mechanism of actin-dependent Golgi organization is not clear, it is conceivable that the first steps of organizing a Golgi ribbon in mammalian cells may depend on the actin cytoskeleton while the later stages of Golgi ribbon formation may involve microtubules.

2.e. Contribution of Lipids to Golgi Structure and Function

The Golgi complex serves as a platform for not just regulating membrane trafficking and sorting events, but is actively involved in mediating mitosis, apoptosis and signaling events. The complex function and architecture of this organelle is maintained by a wide array of molecules that are associated with the Golgi either as residents or as recycling molecules. In addition to proteins, another important class of molecules that maintains Golgi structure and function is lipids.

2.e.1. Importance Of Glycerophospholipids In Golgi Structure And Function

Lipids not only modulate Golgi activities, but many of them are actually synthesized and processed in the Golgi apparatus, thus forming an integral part of the organelle. Some diacylglycerol (DAG) is generated at the Golgi during SM production [125,126] and during phospholipase D mediated hydrolysis of phosphatidylcholine (PC) to produce phosphatidic acid (PA), which is then processed to DAG [127,128,129,130]. DAG serves to recruit TGN vesiculation factor protein kinase D (PKD) to Golgi membranes, and thus functions to modulate cargo trafficking through the organelle [131,132,133]. Thus, DAG and PA, two lipid molecules generated at Golgi membranes are involved in modulating Golgi structure-function. In addition, there is some speculation that generation of phosphatidylinositol phosphate (PIP) and phosphatidylinositol 4,5-bisphosphate (PI(4,5)P2) at the Golgi may promote vesicular fission [133].

Phosphatidylinositol (PI) and its phosphorylated derivatives are important signaling molecules that play roles in regulating Golgi structure and function. Since phosphoinositide production at various cellular locations is tightly regulated, phosphoinositide function is highly localized. Among the different types of phosphoinositides, phosphatidylinositol-4-phosphate (PI4P) is enriched at Golgi membranes. PI4P is mostly synthesized at the Golgi apparatus [134] and is not only important for recruiting components of the trafficking machinery and mediating vesicular transport, but also plays an important role in non-vesicular transport of lipids from the ER directly to the Golgi and in maintainence of membrane contact sites by recruiting lipid transfer proteins (described in detail below) [135,136,137,138,139]. In addition, PI4P has been implicated in vesicular budding and maintainence of Golgi structural integrity through regulation of the actin cytoskeleton [139,140,141].

Since most of the work on the Golgi pool of PI4P was conducted in yeast where a Golgi ribbon is lacking, it would be interesting to know if the mammalian Golgi PI4P pool is involved in additional functions at the Golgi ribbon. Phosphoinositide 4-kinases (PI4K) catalyzes the phosphorylation of PI to PI4P, which then acts as a precursor for synthesis of other PIs, including PI(4,5)P2, PI(3,4)P2 and PI(3,4,5)P3. There are four PI4K enzymes that fall into two groups: type II and type III. While PI4KIIa, PI4KIIb and PI4KIIIb are localized to the Golgi region, PI4KIIIa is not [142,143]. Another important signaling phospholipid is PI(4,5)P2, which is predominantly produced at the plasma membrane, but a small portion is localized at the Golgi [144].

Although PI(4,5)P2 is thought to promote anterograde trafficking in polarized cells, some studies implicate its role in Golgi structural organization due to its interaction with cytoskeletal elements [139,145,146,147,148,149]. PI(4,5)P2 can be generated by phosphoinositide phosphate 5-kinases (PIP5K) but the localization of this enzyme to the Golgi complex is unclear. However, there is evidence of its activity at Golgi membranes upon activation of the small GTPase Arf [142,150]. PI(4,5)P2 can also be generated by phosphoinositide 3-kinases, some isoforms of which are associated with the Golgi complex [142].

Even though much is known about the Golgi localized PI kinases, very little information is present on Golgi localized phosphatases regarding regulation of PI and how they might contribute to Golgi structure and function [142]. Nevertheless, the phosphoinositide phosphatase Sac1, a major regulator of the Golgi pool of PI4P, and inositol polyphosphate 4-phosphatase, inositol polyphosphate 5-phosphatase and oculocerebrorenal-1 with enzymatic activity towards PI(4,5)P2 are among the phosphatases that localize to the Golgi region and regulate Golgi phosphoinositide turnover [139].

2.e.2. Role of Sphingolipids and Lipid Transfer Proteins in Golgi Structure and Function

Another class of lipids that is intimately involved with the Golgi complex and plays an important role in Golgi structure and function is sphingolipids. Sphingolipids are synthesized at the Golgi from their precursor ceramide and are degraded in lysosomes. Ceramide is transported from its site of synthesis in the ER to the Golgi for glucosylceramide (GlcCer), lactosylceramide (LacCer), ganglioside (GM) and sphingomyelin (SM) synthesis. GlcCer is made at the cis-side of the Golgi where GlcCer synthase resides [151], while SM is made at the trans-side of the Golgi where SM synthase resides [151,152]. Thus, sphingolipid synthesis is compartmentalized. Normally, bulk ceramide transport occurs by vesicular trafficking, but when vesicle trafficking from the ER to the Golgi was blocked, sphingolipid synthesis continued to occur [153,154], suggesting the existence of a non-vesicular trafficking mechanism. Non-vesicular routes are thought to exist to not only facilitate rapid and efficient flow of lipids from one region of the cell to another, but to also connect membranes that are not normally linked by vesicular transport, such as the nucleus and mitochondria where sphingolipid metabolism is thought to occur [155]. Many proteins and lipids are involved in partitioning of sphingolipids at the Golgi apparatus to ensure their nonrandom distribution in cells. Membrane contact sites and soluble lipid transfer proteins including ceramide transfer protein (CERT), oxysterol binding protein (OSBP) and four-phosphate adaptor protein 2 (FAPP2) mediate rapid diffusion of lipids from one subcellular region to another.

2.e.2.1. CERT and Stabilization of ER-Golgi Contact Sites

CERT was discovered and characterized by Hanada et al. in a screen for genes that could restore SM levels in the LY-A cell line where traffic of ceramide from the ER to the Golgi was defective [156]. Thus, CERT was found to mediate ATP-dependent trafficking of ceramide from the ER directly to SM synthase at trans-Golgi membranes by bypassing cis- and medial- Golgi compartments. The steroidogenic acute response protein related lipid transfer (START) domain at the C-terminus of CERT is capable of interacting specifically with ceramide and is responsible for extracting and delivering ceramide [156]. This non-vesicular trafficking route is facilitated by the ability of the CERT protein to interact with both ER and Golgi membranes, perhaps simultaneously, at trans-ER-trans-Golgi membrane contact sites [157,158,159,160] via its ER and Golgi interacting domains [156]. Interaction of CERT with the ER occurs via an ER–resident protein called vesicle associated membrane protein associated protein (VAP) while interaction with Golgi membranes occurs by binding of its pleckstrin homology domain to PI4P enriched trans-Golgi membranes [156,157,161,162].

VAP is an ER-resident type II membrane protein and mammalian cells express two VAPs, VAP-A and VAP-B, which can form homodimers and heterodimers [163]. VAP-C is a splicing variant of VAP-B [163]. Recently, VAP-A was shown to co-immunoprecipitate with CERT [164,165], and disruption of the ER-binding domain of CERT inhibited ceramide trafficking in cells [164]. Since VAP not only interacts with CERT, but also with other proteins including SNAREs [157,166,167] and occludin [157,162], it is likely that VAP may not only promote trans-ER-trans-Golgi membrane contact sites, but may also be important for mediating interaction of ER membranes with membranes of other organelles [157]. Since CERT also interacts with PI4P at Golgi membranes, the Golgi pool of PI4P must be regulated for CERT activity. The PI4P pool at the trans-Golgi is maintained by Golgi-localized PI4 kinases IIa, IIIa and IIIb as described above, and CERT activity and targeting to the Golgi region was specifically found to be mediated by PI4 kinase IIIb [143]. CERT activity is also mediated by its phosphorylation status. CERT contains a serine/threonine repeat motif that is subjected to phosphorylation by PKD and possibly casein kinase I [168], and dephosphorylation by protein phosphatase 2 C-ε (PP2Cε) [169]. Phosphorylated forms of

CERT associate less efficiently with Golgi membranes while Golgi association is promoted for non-phosphorylated forms. PKD is known to regulate vesicular trafficking [131,132,133], so it is interesting that PKD also regulates CERT activity and thus non-vesicular trafficking. Intriguingly, overexpression of CERT promotes phosphorylation and activation of PKD [170], presumably by DAG, which is produced by increased conversion of ceramide to SM at the trans-Golgi. This suggests that PKD negatively regulates CERT, whereas CERT positively regulates PKD, and that these two molecules form part of a feedback loop [171]. In addition, PP2Cε is an integral ER membrane protein and is implicated in negatively regulating the stress-activated protein kinase pathway [171,172]. Interestingly, PP2Cε was found to interact with ER resident VAP-A and was shown to dephosphorylate CERT in a VAP-A dependent manner [169], suggesting that CERT function at the Golgi also requires the presence of ER resident integral membrane proteins. Thus, CERT activity at the Golgi is tied to the ER, indicating that CERT may specifically function at ER-trans-Golgi membrane contact sites. Indeed, disruption of Golgi structure by inducing microtubule depolymerization resulted in reduced overlap of CERT with Golgi membranes, with a concomitant decrease in SM synthesis [160].

2.e.2.2. Role of FAPP2 and Transport of GlcCer Within the Golgi Apparatus

Glucosylceramide synthesis occurs on the cytoplasmic leaflet of the cis-Golgi by glucosylceramide synthase. Ceramide required for GlcCer synthesis is delivered from the ER to cis-Golgi largely by vesicular trafficking [152,156]. GlcCer is further processed to glycosylsphingolipids (GSLs), including LacCer and gangliosides in the luminal leaflet of medial/trans-Golgi cisternae. Therefore, GlcCer must be flipped from the cytoplasmic leaflet to the luminal leaflet of Golgi membranes as well as traffic to medial/trans-Golgi membranes. This flipping may be mediated by P-glycoprotein, an ABC transporter (also called MDR1) [173]. The discovery of FAPP2, another soluble lipid transfer protein that specifically interacts with GlcCer via its C-terminal glycolipid transfer protein homology domain and PI4P at Golgi membranes via its PH domain, suggested that FAPP2 could transfer GlcCer from cis-Golgi membranes to medial/trans-Golgi membranes [174,175,176]. However, the direction of GlcCer trafficking is up for debate. While De Matteis and coworkers suggest that FAPP2 mediates GlcCer transport from the cis-Golgi to the trans-Golgi in a non-vesicular manner [177], van Meer's group suggests that GlcCer made at the cis-Golgi is retrogradely trafficked to the ER, where it is flipped to the ER lumen and then transported back to the Golgi to be further glycosylated into GSLs [152]. However, interaction of FAPP2 with ER membranes and definitive identification of the GlcCer flippase have not yet been demonstrated.

2.e.2.3. OSBP, Cholesterol Sensing and Maintainence of ER-Golgi Contact Sites

Cholesterol plays an important role in mammalian cells. It is a major lipid species of the plasma membrane and some intracellular organelles, such as the Golgi and endosomes. Perturbation of membrane cholesterol affects cargo trafficking and signal transduction processes, and therefore its synthesis and distribution must be tightly regulated.

Cholesterol can be synthesized de novo in the ER, but exogenous cholesterol can also be taken up by cells in lipoprotein particles [178]. Cells employ both vesicular and non-vesicular routes to maintain cholesterol homeostasis. It is known that SM and cholesterol levels are coordinately regulated with respect to their metabolism and physical association in membrane rafts [179]. However, the mechanisms that control the coordinated synthesis of cholesterol and SM remain largely uncharacterized.

Oxysterols are produced by the oxidation of cholesterol and are considered to be important signaling molecules as well as regulators of cholesterol homeostasis [179,180,181]. Oxysterols can mediate down regulation of cholesterol biosynthesis and uptake, and promote cholesterol removal and storage [179]. One of the oxysterols, 25-hydroxycholesterol (25-OH) was found to promote translocation of OSBP, a sterol transfer protein, from the ER (where it interacts with ER resident protein VAP-A) to the Golgi region (through its PI4P interacting PH domain) [165,182]. 25-OH treatment also resulted in enhanced SM synthesis by promoting CERT mediated ceramide trafficking from the ER to the trans-Golgi [165,182]. In addition, cholesterol-dependent phosphorylation of OSBP was found to negatively regulate its Golgi localization pattern [183,184], suggesting that sterol levels can also regulate OSBP and CERT activity. OSBP was found to be important for the interaction of CERT with VAP, but OSBP itself showed only a weak or no direct interaction with CERT [165]. Thus CERT function and SM levels can be regulated by sterols, OSBP and VAP [165,183,185]. It is also thought that OSBP and CERT may function in a coordinate manner at the ER-trans-Golgi membrane contact sites via VAP, and mediate cellular cholesterol/SM homeostasis at the Golgi [165].

2.e.2.4. Sphingomyelin and CERT Function in Flies and Mammals

Although OSBP orthologs are found in *S. cerevisiae*, CERT has not been identified in *S. cerevisiae*. In addition, *S. cerevisiae* does not produce SM, but instead synthesizes inositol phosphorylceramide (IPC), a SM analogue. Therefore, it remains unclear whether non-vesicular trafficking of ceramide from the ER to the Golgi occurs in yeast. Perhaps ceramide trafficking in yeast occurs solely by the default bulk flow vesicular pathway. Given that no membrane contact sites have been identified in yeast, it is possible that there is no requirement for a ceramide transfer protein. However, a *Drosophila* CERT does exist [186].

Flies lacking a functional CERT protein display decreased ceramide phosphoethanolamine (CPE) (the SM analogue in *Drosophila*) levels, reduced life span, premature aging, and increased susceptibility to oxidative stress, which could be rescued by introduction of a functional gene [186]. The enzyme that mediates *Drosophila* CPE synthesis is unclear. No homologues of SM synthase exist in flies, except for an SM synthase-related (SMSr) protein [187,188]. Another study shows that SMSr is predominantly present in the ER, not the Golgi, and is capable of only producing small amounts of CPE in flies, while bulk production of CPE is independent of SMSr [189]. The latter study suggests that *Drosophila* cells contain two distinct CPE synthases: one that catalyzes the transfer of the PE head group (like SMSr), and another unrelated enzyme that catalyzes transfer of CDP-ethanolamine for bulk CPE production. The identity and location of the bulk CPE enzyme is not known. Therefore, the location and mechanism of CPE production in flies remains ambiguous. In addition, although parallel pathways can be drawn for IPC and its derivatives in yeast and CPE synthesis in *Drosophila* with that of mammalian SM synthesis, their modes of regulation cannot be directly compared. Besides, *Drosophila* predominantly synthesizes sphingolipids with shorter acyl chains (namely C14 and C16) while mammals generally synthesize sphingolipids with longer acyl chains (at least C18) [186].

Since it is not clear if *Drosophila* cells contain membrane contact sites, and in consideration of the differences in sphingolipid biosynthesis, it is likely that the mechanistic function of CERT in ceramide trafficking may be different than its mammalian counterpart. While ceramide trafficking by CERT in mammals may occur at membrane contact sites, ceramide trafficking by CERT may occur independently of contact sites in *Drosophila*. Regardless, *Drosophila* CERT plays a role in oxidative stress [186] and may point towards a more sophisticated function for CERT in regulating cellular stress at the Golgi ribbon and ER membrane contact sites in mammals.

Interestingly, a study evaluating CERT knockout mice shows that CERT is essential for embryonic survival as no homozygous mice mutant for the *Cert* gene were born [190]. CERT null embryos had growth defects and were estimated to die around embryonic day 11.5 due to abnormalities in organogenesis. The expected decrease in SM levels and the increase in ceramide levels in the ER of CERT null embryos correlated with altered ER morphology and ER stress, and mitochondrial morphology. But analysis of Golgi structural morphology was not conducted. Although the study found an increase in proapoptotic factors Bax and Bak and a decrease in anti-apoptotic proteins Bcl-2 and Bcl-xL (described below in detail), increased cytochrome c release from the mitochondria (one hallmark of apoptosis) was not observed. In addition, the authors saw an increase in activity of some molecules of the cell survival MAPK signaling pathway. Their data also reveal lack of apoptosis as measured by DNA damage. The authors therefore concluded that elimination of CERT does not lead to apoptosis. However, several questions remain. While CERT may be important for development and organogenesis and increased ceramide levels may be detrimental, it is unclear how global ceramide increase over an extended period of time (11.5 days) might modify signaling and organelle morphology, since response to acute ceramide increase was not monitored. In addition, it is not clear why both pro-apoptotic factors and survival factors are elevated when decreased apoptosis and increased cell cycle defects are seen. It is important to mention here that increased ceramide levels have been previously correlated with G0/G1 cell cycle arrest [191] and therefore it would be interesting to analyze how ceramide contributes to both apoptosis and cell cycle signaling events simultaneously. Also, not all apoptotic pathways are mediated by mitochondria and cytochrome c release. Hence it would be important to analyze ceramide levels and signaling molecules prior to day 11.5 in order to determine the progression of events leading to growth defects and embryonic lethality.

3. APOPTOSIS AND GOLGI STRUCTURAL ORGANIZATION

Apoptosis (programmed cell death) is an evolutionarily conserved, highly regulated form of cell suicide that is designed to remove extraneous or damaged cells without causing inflammation [192]. This process is important for multi-cellular organisms to maintain normal growth and development, because dysregulation can lead to disease. Signals that induce cell death can either come from different organelles within the cell such as DNA damage in the nucleus, or can be mediated by extrinsic signals, for example, death receptor ligation [192]. During apoptosis, damaged cells are disassembled and packaged into membrane-bound blebs that signal the neighboring cells to phagocytose them.

3.a. The Cellular Machinery for Apoptosis

Disassembly of apoptotic cells is chiefly mediated by the activity of a family of cysteine proteases called caspases, which cleave after specific aspartate residues within target proteins thereby irreversibly inactivating their function [193]. Caspases are normally synthesized as inactive zymogens but can be activated by cleavage or oligomerization in response to stress [194]. There are at least 12 members within the mammalian caspase family, although not all of them mediate apoptosis. Caspases-2, -3, -6, -7, -8, -9 and -10 are activated during apoptosis, while caspases-1, -4, -5 and -11 are activated during inflammatory responses [195]. In addition, caspase-12, an ER-specific caspase expressed in mice has a role in ER stress (Nakagawa T et al, 2000). Human caspase-12 has no protease activity and may regulate cytokine release in response to bacterial lipopolysaccharide during infection [196]. Caspases that can be activated during apoptosis can be divided into two groups: initiator and effector caspases. Initiator caspases, including caspases-8, -9 and -10, and possibly caspase-2, are upstream caspases as they can initiate downstream apoptosis signaling pathways [195]. Initiator caspases can be activated by self-cleavage after adaptor mediated oligomerization of their characteristically long prodomains. Initiator caspases are generally recruited to large molecular weight complexes that mediate caspase oligomerization by bringing caspase molecules in close proximity to each other [194]. Caspase-8 is activated by its recruitment to the death inducing signaling complex upon ligation of death receptors [195]. Similarly, caspase-9 is activated by the Apaf-1 (apoptosis promoting factor -1) apoptosome that is formed upon release of cytochrome c from mitochondria in response to pro-apoptotic stimuli [195]. Recently, caspase-2 was shown to be activated after recruitment to a high molecular weight complex, the PIDDosome, in the cytoplasm in response to heat shock stress [197]. The PIDDosome complex consists of the death-domain containing protein PIDD (p53 induced protein with a death domain), an adaptor protein RAIDD (RIP associated ICH-/CED-3-homologous protein with a death domain) and caspase-2 [198]. Activated initiator caspases, except caspase-2 (discussed in detail later), cleave and activate effector caspases, including caspases-3, -6 and -7 [194,195]. Effector caspases in turn cleave a small subset of specific cellular proteins, leading to disassembly of intracellular organelles and membrane blebbing.

Many pro- and anti-apoptotic proteins contribute towards regulating the pro-survival and pro-death pathways. Bcl-2 and Bcl-xL are two anti-apoptotic proteins that belong to the Bcl-2 family of proteins that help sequester mitochondrial pro-apoptotic factors and preserve the mitochondrial membrane [199,200]. On the opposite end are Bax and Bak, two pro-apoptotic proteins that promote release of mitochondrial pro-apoptotic factors by compromising the mitochondrial membrane [199,200,201,202]. Furthermore, Bim, another pro-apoptotic member of the Bcl-2 family, is thought to mediate caspase activation and release of Ca^{2+} from the ER [203,204]. Additionally, the inhibitor of apoptosis protein (IAP) family of proteins can either bind directly to the active site of specific caspases and inhibit their activity or function as E3 ligase and target the caspases to the proteasome for degradation [205,206,207,208].

Although different types of intracellular stress may activate one or more of the different initiator caspases, the downstream signaling pathways that disassemble damaged cells usually converge at the level of effector caspases, cellular protein cleavage and intracellular organelle disassembly. Some of the apoptotic machinery that mediates cleavage of cellular targets is found at the Golgi, suggesting that this organelle may be able to sense specific stress signals and regulate downstream signaling pathways. Caspase-2, a unique caspase that possesses structural properties of an initiator caspase [209,210] but the substrate specificity of an effector caspase [193], localizes to both the nucleus and the cytoplasmic face of the Golgi apparatus [211]. Indeed, numerous studies suggest that caspase-2 is activated in the nucleus [212,213,214], while Bouchier-Hayes L et al. [197] clearly show that caspase-2 is activated in punctate spots in the cytoplasm in response to stress. However, it remains to be seen if the punctate spots carrying the PIDDosome-caspase-2 activation complex are formed on disassembled Golgi membranes in response to stress. This is a likely scenario because the disassembled Golgi membranes were found to be associated with endogenous caspase-2 [197]. The dual localization pattern of caspase-2 suggests that the nuclear pool of caspase-2 could mediate apoptosis after DNA damage whereas the Golgi pool could mediate apoptosis in response to stress signals sensed by the secretory pathway, particularly the Golgi apparatus itself [215]. In fact, several known substrates of caspase-2 are localized at the Golgi, including golgin-160 [211] and giantin [216]. Golgin-160 is cleaved by capspases-2, -3 and -7, but the cleavage by caspase-2 is rapid and precedes cleavage by caspase-3 and Golgi disassembly, thus suggesting that caspase-2 activation at the Golgi is an early event [211]. The exact mechanism of caspase-2 activation is not well understood. Nevertheless, Golgi disassembly and apoptosis was delayed in cells expressing a non-cleavable golgin-160 (where all three caspase sites were mutated) after certain apoptotic stimuli at a step upstream of caspase-2 activation [217].

BIR repeat containing ubiquitin-conjugating enzyme (BRUCE), a member of the IAP family of proteins is localized in the Golgi region [218] and has been shown to possess anti-apoptotic properties in mammalian cells [219,220]. Thus, BRUCE may function to negatively regulate caspase-2 activity at the Golgi [221]. In addition, many proteins of the death receptor family of proteins, including the tumor necrosis factor receptor-1 (TNFR-1) and Fas are mostly localized to the Golgi region at steady state [222,223], suggesting that there is precedence for the proposal that the Golgi apparatus can sense cellular stress and integrate downstream signaling pathways. Recently, BRUCE has been implicated in mitosis, specifically during cytokinesis where it interacts with other proteins involved in mitosis [224]. Thus, BRUCE may have multiple roles in apoptosis and in mitosis.

Stress signals that might be sensed by the secretory pathway include perturbation of vesicular and non-vesicular trafficking, perturbation of Golgi structure and infection with certain enveloped viruses [215]. Disassembly of Golgi structure during apoptosis is similar to mitotic disassembly, but unlike mitosis, apoptotic disassembly is an irreversible process [225]. During apoptosis, the lateral connections between Golgi stacks that mediate the Golgi ribbon are lost. Golgi stacks are dispersed in the cytoplasm and do not maintain contact with the MTOC, while the cisternae are further broken down to form dispersed vesicles and tubules. Presumably, during the process of Golgi disassembly, trans-ER-trans Golgi contact sites are also disassembled. Supporting this, treatment of cells with tumor necrosis factor α (TNFα) shows reduced overlap of CERT with Golgi membranes (Chandran, unpublished). This process of disassembly is mediated by cleavage of Golgi localized proteins such as golgin-160, GRASP-65, p115, GM130, giantin, the t-SNARE syntaxin-5, the intermediate chain of dynein and the p150 (glued) subunit of dynactin [211,216,226,227,228,229,230], which are known to contribute to Golgi architecture and structural maintenance. The delay in Golgi disassembly upon expression of caspase resistant mutants of golgin-160 [211,217], GRASP-65 [226] and p115 [227] further underlines the importance of these proteins in Golgi structure. However, several important questions regarding the model that the Golgi apparatus is a stress sensor remain. One big question is what is the role of Golgi disassembly during apoptosis? One thought is that Golgi disassembly may play a role in signal transduction events that feed into the apoptotic signaling cascade. Interestingly, caspase mediated cleavage of some of the Golgi structural proteins including golgin-160 and p115 result in fragments with an exposed nuclear localization signal (NLS) [227,231]. Ectopic expression of these fragments (which contain sequence motifs of transcriptional regulators) were targeted to the nucleus, suggesting that these fragments can impact the apoptotic process by modulating gene expression [227,231]. Although over expression of caspase cleavage fragments of golgin-160 did not promote Golgi fragmentation, over expression of p115 fragments did result in Golgi disassembly and apoptosis [227,231]. Thus, caspase cleavage fragments of some Golgi structural proteins may play an active role in regulating apoptosis, supporting the view that Golgi structure may serve as a platform to sense and integrate stress signals. But what genes the nuclear localized fragments of Golgi structural proteins might modulate is an important question that still remains completely uncharacterized. Another question of significance that remains to be answered is whether Golgi disassembly is an effect of the apoptotic process or whether Golgi disassembly itself is a signal for apoptosis [215]. Regardless, it is clear that Golgi ribbon structure, including ER-trans-Golgi contact sites, is exquisitely sensitive to perturbation, making this organelle an ideal platform to sense cellular stress and integrate signals that determine cell survival or cell death.

Golgi structural proteins are also important in maintaining Golgi structure and organization in *Drosophila*. Depletion of *Drosophila* Golgi structural proteins (GRASP, GM130, and p115) led to collapse of Golgi stacks into vesicular/tubular clusters [52,232]. However, forward trafficking remained unaffected [52,232]. It remains to be seen if *Drosophila* GRASP and p115 undergo caspase-mediated cleavage to generate nuclear targeted fragments and how they might function in the apoptotic signaling pathway. In addition, although orthologues for caspases-3, -7 and -9 exist in *Drosophila*, there is no indication of presence of a Golgi localized caspase-2. Examination of some of these issues may help determine whether cleavage of Golgi proteins and Golgi disassembly play an active role in the apoptotic process both in flies and humans.

3.b. Ceramide Mediated Apoptosis And Golgi Ribbon Architecture

Stress signals that must be sensed by the secretory pathway not only include disassembly of Golgi structure and perturbation of vesicular and non-vesicular trafficking by cleavage of Golgi structural proteins, but also include dysregulation of cellular lipid homeostasis. While membrane sterol regulation is well characterized [178], mechanisms that regulate ceramide and sphingolipid levels in cells is less understood. As described above, ceramide serves as the backbone for sphingolipid biosynthesis, but it also acts as an important second messenger in cells [233]. Therefore, biosynthesis and cellular distribution of ceramide must be tightly regulated. In order to appreciate how ceramide contributes to lipid homeostasis, signaling, Golgi structure and secretory trafficking, and cellular processes such as growth, proliferation and apoptosis, it is important to first understand how ceramide is synthesized and how it generates other important signaling lipids including sphingosine, sphingosine-1-phosphate and ceramide-1-phosphate [234,235].

3.b.1. Ceramide Biosynthesis and Metabolism

Ceramide can be synthesized either de novo or by the hydrolysis of sphingolipids, particularly SM (Figure 2). The de novo synthesis of ceramide occurs in the ER by the condensation of serine and palmitoyl Co-A to generate 3-ketodihydrosphingosine [236]. 3-ketodihydrosphingosine is then reduced to form dihydrosphingosine (also called sphinganine), which is subsequently N-acylated to produce dihydroceramide by the action of ceramide synthase. There are six isoforms of ceramide synthase in mammals that show specificity for the acyl chain length [236]. Dihydroceramide is later desaturated to produce ceramide [236]. The ceramide formed in the ER can be trafficked to various locations in the cell by vesicular and CERT-mediated-non-vesicular routes, and be converted to other sphingolipids. In cultured cells, ceramide is mostly trafficked from the ER to the trans-Golgi via CERT. There it is converted to SM by the action of SM synthase 1, where the phosphatidylcholine (PC) head group is transferred onto the ceramide backbone generating DAG, which is an important lipid in maintaining Golgi structure and function, as described above [131,156,237]. Ceramide can also be converted to glucosylceramide (GlcCer) by glucosylceramide synthase after delivery of ceramide to cis-Golgi membranes by vesicular transport [152,156,236]. GlcCer can further be glycosylated to other GSLs through the activity of another lipid transfer protein, FAPP2 [152,176,177].

Furthermore, ceramide can also be converted to sphingosine, one of the other important signaling molecules, by the action of ceramidases. Three types of ceramidases (acid, neutral and alkaline) are localized at different regions of the cell reflecting their pH optima for activity [238]. Sphingosine can either act by itself to regulate signaling pathways or can be phosphorylated to sphingosine-1-phosphate, another important signaling lipid that promotes cell growth and proliferation. Alternatively, sphingosine can be converted to ceramide. Sphingosine-1-phosphate can be degraded by sphingosine-1-phosphate lyase or converted back to sphingosine by a phosphatase [97,236,239,240]. In addition to serving as a precursor molecule for sphingolipids, ceramide can also be phosphorylated by ceramide kinase to generate ceramide-1-phosphate, which can be recycled back by a phosphatase [236,241].

In addition to biosynthesis, ceramide can also be generated by sphingomyelinase (SMase) mediated hydrolysis of SM. Although ceramide is synthesized de novo at the ER, ceramide production by SM hydrolysis can occur at various locations in the cell. Different SMases that show different subcellular localization based on their pH optima for acivity: lysosomal acid SMase, zinc-dependent secretory SMase, neutral magnesium-dependent SMase and alkaline SMase [236,242].Hydrolysis of GSLs yields GlcCer and galactosylceramide (GalCer), which can then be hydrolyzed by β-glucosidases and galactosidases to release ceramide [236,243].

Thus, SM and GSLs that are produced from ceramide can be transported to the plasma membrane by vesicular trafficking where they can be subsequently metabolized back to ceramide. SM and GSLs can also be circulated through the secretory and endosomal pathway where they can be degraded to ceramide and sphingosine in the lysosome. Since sphingosine is fairly soluble, it can leave the lysosome and diffuse to various locations in the cell, including the ER, where it can re-enter the ceramide cycle [236].

Ceramide and its derivatives including sphingosine, sphingosine-1-phosphate, and ceramide-1-phosphate are important signaling lipids involved in regulating downstream targets that mediate various cellular functions. Ceramide regulates various aspects of cell growth, differentiation, proliferation, necrosis and apoptosis [233,236,244,245,246]. Ceramide is known to activate protein phosphatases PP1A and PP2A, protein kinase C, raf-1, and the kinase-suppressor of Ras [236,247,248,249,250]. Ceramide-1-phosphate activates DNA synthesis and cell division [251]. It also mediates inflammation by activating phospholipase A_2 and stimulating the release of arachidonic acid [236,252]. Ceramide-1-phosphate also blocks apoptosis by inhibiting acid sphingomyelinase (reviewed in [251]). Sphingosine is known to induce cell cycle arrest and apoptosis by activating protein kinase A [236,253,254]. Sphingosine-1-phosphate interacts with the G-protein coupled sphingosine-1-phosphate receptors at the plasma membrane and positively regulates G-protein signaling involved in cell growth, proliferation, migration, inflammation, angiogenesis, vasculogenesis, cell survival. Conversely, sphingosine-1-phosphate negatively regulates apoptosis [236] by activating ERK signaling cascade, Ras, PI3K, phospholipase C, and p125^{FAK} to name a few [255]. Therefore, ceramide and its derivatives play an important role in modulating cellular processes that ultimately involve cell survival and cell death decisions.

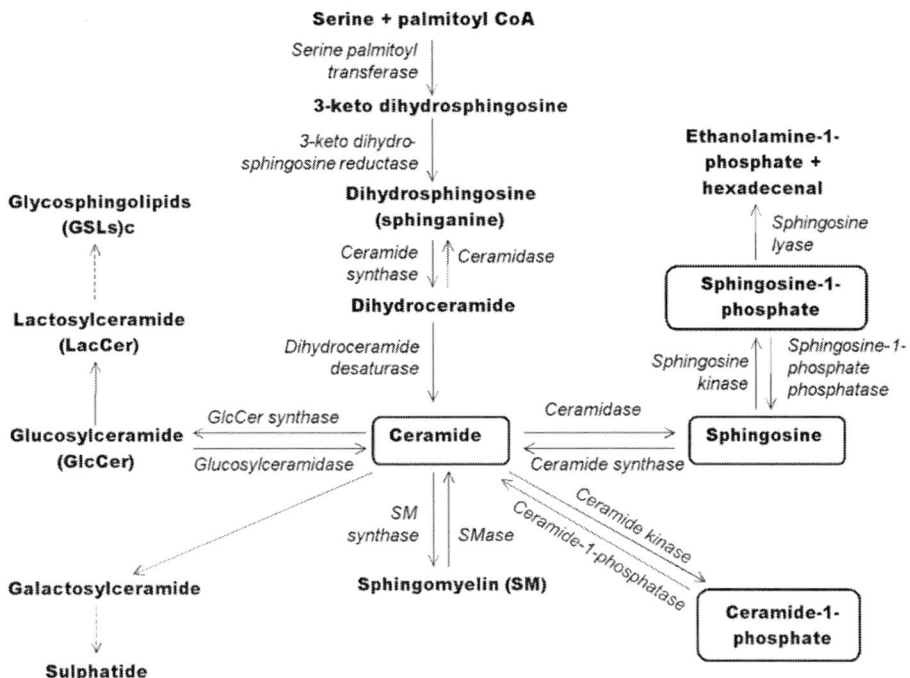

Figure 2. Sphingolipid metabolism pathways. Enzymes that catalyze the reactions are italicized while the substrates and products are in bold. The boxes represent sphingolipids with signaling roles, as described.

3.b.2. Ceramide Regulation and Golgi Structure

Since ceramide is central to sphingolipid biosynthesis, and ceramide and sphingolipid synthesis is closely tied into the secretory pathway, it is likely that perturbation of the secretory pathway could lead to dysregulation of ceramide homeostasis and its downstream functions. Since the ceramide concentration in cells is maintained at a higher concentration than sphingosine and sphingosine-1-phosphate, one can envision that small changes in ceramide levels could significantly increase sphingosine or sphingosine-1-phosphate levels [189,236]. Similarly, blocking conversion of ceramide to SM, GlcCer, or sphingosine could result in accumulation of ceramide in the ER and other regions of the cell. Thus, cells have to be able to sense and regulate ceramide levels. The recent identification of SM synthase 1 related enzyme (SMSr) suggests that this protein may be able to sense ceramide levels in the ER and regulate ceramide homeostasis [189]. Unlike SM synthase I, SMSr is predominantly found in the ER and catalyzes the synthesis of the SM analog EPC in the ER lumen, probably by promoting flipping of ceramide from the cytosolic face of the ER to the active site of the enzyme in the ER lumen. Surprisingly, SMSr produces only minor amounts of EPC, about 300 times less than SM produced by SM synthase. This can perhaps be attributed to the low dissociation rate of SMSr from its product EPC. Also, CERT-mediated removal of ceramide present on the cytosolic side of the ER under normal conditions could account for low levels of cellular EPC. Interestingly, depletion of SMSr by RNA interference resulted in a drastic increase in ceramide levels in the ER, and simultaneous disassembly of ER exit sites and perturbation of the Golgi apparatus. ER exit sites or transitional ER elements are thought to arise from ribosome-free ER membrane patches where COPII mediated vesicles emerge [45,256]. Vacaru et al. show that the fragmentation of early secretory pathway components when SMSr is inhibited is due to the accumulation of ceramide in the ER [189]. However, an increase in ER ceramide levels could likely cause the entire pool of SMSr enzyme to be engaged with EPC, with many of the ER lumen ceramide molecules remaining in an unbound state. It is possible that the EPC-bound SMSr signals an attenuation of ceramide biosynthesis, and/or an increase in ceramide degradation. Thus, the coordinated action of SMSr and CERT may help maintain ceramide homeostasis and structural components of the secretory pathway, including the Golgi apparatus.

3.b.3. Ceramide and Apoptosis

We know that inhibition of SMSr mediated ceramide regulation in cells is concurrent with ceramide accumulation in the ER [189]. However, it is not clear whether SMSr has to be inactivated during apoptosis for ceramide levels to increase at the Golgi and ER, although this would be a likely scenario. The ceramide-apoptosis link is best described in studies where TNFα was used as an inducer of pro-apoptotic stress. Rapid ceramide generation at the plasma membrane after TNFα treatment is generally dependent on activation of early but not late caspases, which activate sphingomyelinase and thus increase cellular ceramide [257,258]. Inhibition of caspase-8 is associated with decreased levels of ceramide accumulation in TNFα-treated cells and consequently decreased apoptosis, while inhibition of downstream caspases was shown to have no effect on ceramide levels, indicating that ceramide generation is often upstream of effector caspases [257,259]. However, whether caspase-8 acts upstream or downstream of anti-apoptotic factor Bcl-2 remains unclear [257,259,260].

The effect of excess ceramide on Golgi organization is not well understood. Since Golgi biogenesis is closely linked to ER exit sites [45,256], accumulation of ceramide in the ER could lead to perturbation of proteins involved in ER exit site maintenance or components of COPII machinery and could potentially destabilize ER exit site morphology, thereby resulting in Golgi disassembly [189]. On the other hand, excess ceramide levels may promote apoptotic signaling and induce Golgi structure disassembly by activating caspase-mediated cleavage of Golgi structural proteins. But the mechanism of how increased ceramide levels at different cellular membranes activate the apoptotic signaling cascade needs to be examined. For example, it would be interesting to understand how normal ceramide levels in cells do not lead to activation of the apoptosis signaling cascade, but an increase in ceramide concentration is associated with activation of caspases. SMSr, described above, could serve as a sensor for ceramide levels at the ER. But what about ceramide generated at the plasma membrane or other subcellular organelles and trafficked to other regions of the cell by the vesicular and non-vesicular routes? How are these ceramide concentrations sensed and regulated? For example, ceramide can be generated from SM in endosomal/lysosomal compartments and at the plasma membrane by acidic and neutral sphingomyelinases (aSMase and nSMase), respectively. Both nSMase and aSMase activities are known to be stimulated by apoptosis, particularly by activation of death receptors TNFR-1 and Fas [261,262,263]. Not much information exists on activation of aSMase by TNFα, but it may be mediated by specific adaptor proteins [261]. Similarly, little is known about the downstream targets of ceramide generated by aSMase, although there is some implication of involvement of PKCζ, c-Jun amino terminal kinase (JNK), and caspases [261]. On the other hand, nSMase activity can also be mediated by a protein called factor associated with nSMase activation (FAD) in response to cellular stress [263,264]. In addition, nSMase activity is regulated by oxidative stress, including that generated by hydrogen peroxide treatment [263], although it is unclear if reactive oxygen species (ROS) generate ceramide, or ceramide-mediated release of cytochrome c from the mitochondria generates ROS. Ceramide generated by nSMase is known to activate ceramide-activated protein kinase by enhancing its autophosphorylation, which eventually leads to activation of the extracellular signal-regulated kinase (ERK) signaling cascade [261,265]. Therefore, activation of aSMase and nSMase seem to follow different pathways and it remains to be seen if activation of a particular SMase occurs in response to a particular stress [263]. TNFα may also stimulate the de novo synthesis pathway of ceramide, although this is not well characterized [257].

Ceramide has also been implicated in mitochondrial-mediated apoptosis. Activation of aSMase leads to lysosomal generation of ceramide, where it binds cathepsin-D and is translocated to the mitochondria by an unknown mechanism resulting in release of cytochrome c [257]. Release of cytochrome c could occur through ceramide channels that form on the outer membrane of mitochondria upon ceramide-mediated membrane permeabilization [266]. Formation of ceramide channels was found to be negatively regulated by Bcl-xL, an anti-apoptotic protein of the Bcl-2 family [266]. But the mechanism of ceramide delivery to mitochondrial membranes and how it mediates permeabilization of mitochondria membranes remains to be elucidated.

In view of the fact that ceramide is generated not only at the ER, but also at various subcellular regions that are involved in membrane traffic, perhaps the components of the secretory pathway might sense cellular ceramide levels and respond appropriately. Since several studies report collapse of the early secretory pathway, including the Golgi apparatus, upon ceramide accumulation in response to cellular stress [189,267,268], it seems reasonable to speculate that the Golgi may serve to function as a platform for sensing stress and integrating apoptosis signals. Indeed the choice of the Golgi apparatus for a sensor of cellular stress seems ideal because it sits at the center of the secretory pathway and many components of the apoptotic and signaling machinery, including caspase-2 are found there. Therefore it seems likely that ceramide-induced apoptosis could be accompanied by caspase-mediated cleavage of Golgi structural proteins that contribute to collapse of the Golgi complex and perhaps further amplification of the apoptotic signal via the nuclear targeted Golgi protein fragments. Interestingly, a recent study found that the trans-Golgi localized SM synthase-1 enzyme that catalyzes the conversion of ceramide to SM, is also a substrate for caspases, including caspase-2, during pro-apoptotic signaling events [269]. Given that CERT is associated with Golgi membranes at the ER-trans-Golgi membrane contact sites while delivering ceramide to SM synthase 1, CERT and SM synthase 1 can be considered as molecules that help maintain reduced ceramide levels at the ER and the Golgi complex. It is interesting to speculate that CERT could act as a sensor of stress and mediate signaling. In vitro studies suggest that CERT is as a substrate for caspase-2 (Chandran et al, unpublished data), although CERT cleavage in cells in response to pro-apoptotic stress is currently being assessed. But with the finding that induction of apoptosis in human cells lines upon TNFα treatment resulted in increased transcription of Good pasture binding protein (GPBP), a functional splice variant of CERT, the concept that CERT might sense and integrate cellular stress is not unprecedented [270]. The identification of TATA-like and NFkB-like elements within the promoter region of the GPBP gene suggests that GPBP and presumably CERT is responsive to TNFα [270]. Although the study examined GPBP mRNA levels in the presence and absence of TNFα, GPBP protein levels were not monitored. Thus, the mechanism by which cells handle excess GPBP/CERT remains unclear. While this study only looked at GPBP and its role in immunity, it potentially positions CERT within the apoptotic signaling pathway. Intriguingly, the corresponding promoter in mice lacks the NFκB-like element and shows no transcriptional activity in response to TNFα [270]. Furthermore, CERT protein has a putative NLS whose role has not yet been identified [271].

It is conceivable that CERT cleavage in vivo in response to cellular stress, particularly that induced by TNFα, could result in a nuclear targeted fragment (similar to the scenario seen for golgin-160 and p115 [227,231]) that could possibly modulate the apoptotic signal. In addition, induction of either oxidative stress or UV stress led to homotrimerization and inactivation of the CERT protein, resulting in increased ceramide levels and apoptosis [272]. It is uncertain whether homotrimerization of the CERT protein signals apoptosis or whether increased ceramide levels cause apoptosis.

Indeed, the next big question in ceramide-mediated apoptosis is whether Golgi disassembly is the cause or effect of ceramide accumulation. Do increased ceramide levels cause caspase-mediated cleavage of golgins and result in Golgi structural collapse? Or do the golgins first undergo cleavage and disassemble the Golgi apparatus before ceramide accumulation? The question is whether Golgi disassembly occurs upstream or downstream of ceramide accumulation. It is also possible that Golgi disassembly is a multi-step process, and that some of the steps might occur early during the apoptotic pathway leading to ceramide accumulation. This in turn could induce cleavage of other golgins leading to complete collapse of the organelle.

For example, hypothetically, cellular stress sensed at the Golgi by Golgi localized caspase-2 could result in activation and cleavage of some golgins and GRASP proteins, resulting in subtle changes in Golgi structure and organization, such as loss of connections that maintain the Golgi ribbon and ER-trans-Golgi contact sites. Concurrently, CERT inactivation, either due to loss of contact sites or caspase-2 cleavage would result in inhibition of ceramide consumption at the ER leading to ceramide accumulation at different subcellular regions within the secretory and endocytic pathways.

Although not much is known about the regulation and downstream signaling targets of SMSr, it is possible that either excess ceramide or caspases could inhibit SMSr function in sensing ceramide levels at the ER. Excess ceramide accumulation could trigger amplification of the apoptotic signaling pathway resulting in cleavage of additional golgins and Golgi-associated proteins, resulting in unstacking and vesiculation of the Golgi complex.

4. CELLULAR LOCATION OF THE GOLGI APPARATUS IN STRESS SENSING

It is interesting to speculate that the central cellular location of the Golgi apparatus participates in sensing cellular stress and transducing appropriate downstream signals. As described previously, the microtubule cytoskeleton holds the Golgi in a juxtanuclear region, next to the centrosome.

But why does the Golgi exist in a juxtanuclear region? Does the Golgi exist next to the nucleus only because of its interactions with the centrosome and microtubules, or does the location of the Golgi apparatus influence its functions? Membrane trafficking events and modification of secretory proteins are not affected by perturbation of the peri-centrosomal location of the Golgi [77,78,89], suggesting that the central positioning of the Golgi apparatus in cells may serve other functions. It is possible that the central location of the Golgi can sense subtle changes in the centrosomes or in microtubule polymerization/depolymerization kinetics in response to signaling or cellular events, including mitosis, migration and apoptosis.

It is also possible that the centrally located Golgi apparatus can detect subtle changes in the structure of other organelles through membrane contact sites and respond by promoting cleavage of Golgi structural proteins. Reorganization of the Golgi apparatus due to cleavage of some golgins is known to generate nuclear targeted fragments of those golgins, which are thought to promote apoptosis via gene regulation in the nucleus. The functional relationship between the adjacently located centrosome and Golgi apparatus has been reviewed by Sutterlin et al. [36] in detail. Studies where the Golgi ribbon was perturbed by depletion of golgin-160 or GMAP210 reported loss of cell polarity, directed secretion, cell migration and repair in response to wounding, in a microtubule independent manner [78]. Although Golgi morphology was rescued in this study by using an RNA interference resistant version of golgin-160, rescue of cell polarization and migration phenotypes were not examined. It would be interesting to determine if loss of cell polarity and migration are indeed due to Golgi structural perturbation, or whether the phenotype is due to loss of golgin-160's function in promoting efficient delivery of a subset of cargo molecules to the plasma membrane [79] that could influence cell polarity.

Several Golgi associated proteins have been implicated in maintainence of centrosome structure and function including GM130, and IFT20 and Rab8 GTPase during ciliogenesis [36,273,274,275,276]. Thus it is likely that the Golgi and centrosome are functionally linked and may modulate each other's functions. It is likely that the central location of the Golgi apparatus positions the organelle to sense stress by its interaction with other organelles and transduce signaling events leading to cell cycle, migration, polarity and apoptosis to the nucleus. Since most *Drosophila* cells already exhibit dispersed Golgi stacks whose organization is independent of microtubules and membrane contact sites, it remains to be seen how these cells sense stress within the secretory pathway and cope with apoptosis.

CONCLUSION

From the numerous studies that have been conducted on the Golgi over the past several decades, it can be concluded that the organization of the mammalian Golgi apparatus is complex. The mammalian Golgi ribbon is not only composed of cisternal membranes arranged into stacks, but is also ordered into a continuous ribbon structure that is centrally positioned in the cell and interacts with ER membranes, centrosomes and cytoskeletal elements. But why is the organization of the mammalian Golgi so complex? Although the Golgi of lower organisms such as *S. cerevisiae* and *Drosophila* lack this level of organization, they are still capable of efficiently processing and trafficking cargo. Here, we have speculated that the complex architecture of the mammalian Golgi and its central location sets up the organelle as an ideal platform to sense subtle changes in various parts of the cell during cellular stress and to amplify downstream signaling. Despite the many unresolved questions, there is a vast amount of evidence that implicates Golgi structure-function in apoptotic signaling pathways. Since apoptosis and mitosis share some common elements, particularly disassembly of the Golgi ribbon, it is tempting to speculate that the Golgi ribbon may have additional functions in cell cycle regulation. Interestingly, proteins involved in apoptosis, including BRUCE [224] and nuclear pool of caspase-2 also have additional functions in mitosis. The recent implication of nuclear caspase-2 in the G2/M cell cycle checkpoint [214] further supports an apoptosis-mitosis link and confers novel functions for the Golgi in cell cycle regulation. A more comprehensive approach investigating all molecules involved in maintainence and regulation of Golgi structure, including proteins and lipids, is needed. This could not only provide additional clues in further addressing the Golgi-apoptosis connection, but also reveal new functions for the Golgi ribbon. Determining the order in which apoptosis signaling, ceramide accumulation and Golgi disassembly occurs are future challenges for understanding how the Golgi apparatus contributes to cellular stress sensing and integration of apoptosis signaling.

REFERENCES

[1] Quinn P, Griffiths G, Warren G (1983) Dissection of the Golgi complex. II. Density separation of specific Golgi functions in virally infected cells treated with monensin. *J. Cell Biol.* 96: 851-856.

[2] Tartakoff A, Vassalli P (1979) Plasma cell immunoglobulin M molecules. Their biosynthesis, assembly, and intracellular transport. *J. Cell Biol.* 83: 284-299.

[3] Tartakoff AM (1986) Temperature and energy dependence of secretory protein transport in the exocrine pancreas. EMBO J 5: 1477-1482.

[4] Lagunoff D, Wan H (1974) Temperature dependence of mast cell histamine secretion. *J. Cell Biol.* 61: 809-811.

[5] Holmes KV, Doller EW, Sturman LS (1981) Tunicamycin resistant glycosylation of coronavirus glycoprotein: demonstration of a novel type of viral glycoprotein. *Virology* 115: 334-344.

[6] Saraste J, Kuismanen E (1984) Pre- and post-Golgi vacuoles operate in the transport of Semliki Forest virus membrane glycoproteins to the cell surface. *Cell* 38: 535-549.

[7] Griffiths G, Pfeiffer S, Simons K, Matlin K (1985) Exit of newly synthesized membrane proteins from the trans cisterna of the Golgi complex to the plasma membrane. *J. Cell Biol.* 101: 949-964.

[8] Fries E, Lindstrom I (1986) The effects of low temperatures on intracellular transport of newly synthesized albumin and haptoglobin in rat hepatocytes. *Biochem. J.* 237: 33-39.

[9] Bannykh SI, Rowe T, Balch WE (1996) The organization of endoplasmic reticulum export complexes. *J. Cell Biol.* 135: 19-35.

[10] Presley JF, Cole NB, Schroer TA, Hirschberg K, Zaal KJ, et al. (1997) ER-to-Golgi transport visualized in living cells. *Nature* 389: 81-85.

[11] Hauri HP, Schweizer A (1992) The endoplasmic reticulum-Golgi intermediate compartment. *Curr. Opin. Cell Biol.* 4: 600-608.

[12] Saraste J, Goud B (2007) Functional symmetry of endomembranes. *Mol. Biol. Cell* 18: 1430-1436.

[13] Roth J (1987) Subcellular organization of glycosylation in mammalian cells. *Biochim. Biophys. Acta* 906: 405-436.

[14] Dunphy WG, Rothman JE (1983) Compartmentation of asparagine-linked oligosaccharide processing in the Golgi apparatus. *J. Cell Biol.* 97: 270-275.

[15] Jamieson JD, Palade GE (1971) Synthesis, intracellular transport, and discharge of secretory proteins in stimulated pancreatic exocrine cells. *J. Cell Biol.* 50: 135-158.

[16] Farquhar MG, Palade GE (1981) The Golgi apparatus (complex)-(1954-1981)-from artifact to center stage. *J. Cell Biol.* 91: 77s-103s.

[17] Tamaki H, Yamashina S (2002) The stack of the golgi apparatus. *Arch. Histol. Cytol.* 65: 209-218.

[18] Morre DJ, Ovtracht L (1977) Dynamics of the Golgi apparatus: membrane differentiation and membrane flow. *Int. Rev. Cytol. Suppl.* 61-188.

[19] Brown RM, Jr., Franke WW, Kleinig H, Falk H, Sitte P (1970) Scale formation in chrysophycean algae. I. Cellulosic and noncellulosic wall components made by the Golgi apparatus. *J. Cell Biol.* 45: 246-271.

[20] Pelham HR, Rothman JE (2000) The debate about transport in the Golgi--two sides of the same coin? *Cell* 102: 713-719.

[21] Glick BS, Nakano A (2009) Membrane traffic within the Golgi apparatus. *Annu. Rev. Cell Dev. Biol.* 25: 113-132.

[22] Mowbrey K, Dacks JB (2009) Evolution and diversity of the Golgi body. *FEBS Lett.* 583: 3738-3745.

[23] Kondylis V, Rabouille C (2009) The Golgi apparatus: lessons from Drosophila. *FEBS Lett.* 583: 3827-3838.

[24] Preuss D, Mulholland J, Franzusoff A, Segev N, Botstein D (1992) Characterization of the Saccharomyces Golgi complex through the cell cycle by immunoelectron microscopy. *Mol. Biol. Cell* 3: 789-803.

[25] Rambourg A, Clermont Y (1990) Three-dimensional electron microscopy: structure of the Golgi apparatus. *Eur J. Cell Biol.* 51: 189-200.

[26] Rambourg A, Clermont Y, Hermo L (1979) Three-dimensional architecture of the golgi apparatus in Sertoli cells of the rat. *Am. J. Anat.* 154: 455-476.

[27] Mogelsvang S, Marsh BJ, Ladinsky MS, Howell KE (2004) Predicting function from structure: 3D structure studies of the mammalian Golgi complex. *Traffic* 5: 338-345.

[28] Ladinsky MS, Mastronarde DN, McIntosh JR, Howell KE, Staehelin LA (1999) Golgi structure in three dimensions: functional insights from the normal rat kidney cell. *J. Cell Biol.* 144: 1135-1149.

[29] Ladinsky MS, Wu CC, McIntosh S, McIntosh JR, Howell KE (2002) Structure of the Golgi and distribution of reporter molecules at 20 degrees C reveals the complexity of the exit compartments. *Mol. Biol. Cell* 13: 2810-2825.

[30] Marsh BJ, Mastronarde DN, McIntosh JR, Howell KE (2001) Structural evidence for multiple transport mechanisms through the Golgi in the pancreatic beta-cell line, HIT-T15. *Biochem. Soc. Trans* 29: 461-467.

[31] Marsh BJ, Volkmann N, McIntosh JR, Howell KE (2004) Direct continuities between cisternae at different levels of the Golgi complex in glucose-stimulated mouse islet beta cells. *Proc. Natl. Acad. Sci. U S A* 101: 5565-5570.

[32] Marsh BJ, Mastronarde DN, Buttle KF, Howell KE, McIntosh JR (2001) Organellar relationships in the Golgi region of the pancreatic beta cell line, HIT-T15, visualized by high resolution electron tomography. *Proc. Natl. Acad. Sci. U S A* 98: 2399-2406.

[33] Griffiths G, Simons K (1986) The trans Golgi network: sorting at the exit site of the Golgi complex. *Science* 234: 438-443.

[34] Burke B, Griffiths G, Reggio H, Louvard D, Warren G (1982) A monoclonal antibody against a 135-K Golgi membrane protein. *EMBO J.* 1: 1621-1628.

[35] Colanzi A, Suetterlin C, Malhotra V (2003) Cell-cycle-specific Golgi fragmentation: how and why? *Curr. Opin. Cell Biol.* 15: 462-467.

[36] Sutterlin C, Colanzi A (2010) The Golgi and the centrosome: building a functional partnership. *J. Cell Biol.* 188: 621-628.

[37] Wei JH, Seemann J (2009) Mitotic division of the mammalian Golgi apparatus. *Semin. Cell Dev. Biol.* 20: 810-816.

[38] Persico A, Cervigni RI, Barretta ML, Colanzi A (2009) Mitotic inheritance of the Golgi complex. *FEBS Lett.* 583: 3857-3862.

[39] Sutterlin C, Hsu P, Mallabiabarrena A, Malhotra V (2002) Fragmentation and dispersal of the pericentriolar Golgi complex is required for entry into mitosis in mammalian cells. *Cell* 109: 359-369.

[40] Colanzi A, Hidalgo Carcedo C, Persico A, Cericola C, Turacchio G, et al. (2007) The Golgi mitotic checkpoint is controlled by BARS-dependent fission of the Golgi ribbon into separate stacks in G2. *EMBO J.* 26: 2465-2476.

[41] Feinstein TN, Linstedt AD (2007) Mitogen-activated protein kinase kinase 1-dependent Golgi unlinking occurs in G2 phase and promotes the G2/M cell cycle transition. *Mol. Biol. Cell* 18: 594-604.

[42] Novikoff PM, Novikoff AB, Quintana N, Hauw JJ (1971) Golgi apparatus, GERL, and lysosomes of neurons in rat dorsal root ganglia, studied by thick section and thin section cytochemistry. *J. Cell Biol.* 50: 859-886.

[43] De Matteis MA, Luini A (2008) Exiting the Golgi complex. *Nat. Rev. Mol. Cell Biol.* 9: 273-284.

[44] Levine T, Loewen C (2006) Inter-organelle membrane contact sites: through a glass, darkly. *Curr. Opin. Cell Biol.* 18: 371-378.
[45] Bevis BJ, Hammond AT, Reinke CA, Glick BS (2002) De novo formation of transitional ER sites and Golgi structures in Pichia pastoris. *Nat. Cell Biol.* 4: 750-756.
[46] Seemann J, Jokitalo E, Pypaert M, Warren G (2000) Matrix proteins can generate the higher order architecture of the Golgi apparatus. *Nature* 407: 1022-1026.
[47] Pfeffer SR (2001) Constructing a Golgi complex. *J. Cell Biol.* 155: 873-875.
[48] Shorter J, Watson R, Giannakou ME, Clarke M, Warren G, et al. (1999) GRASP55, a second mammalian GRASP protein involved in the stacking of Golgi cisternae in a cell-free system. *EMBO J.* 18: 4949-4960.
[49] Barr FA, Puype M, Vandekerckhove J, Warren G (1997) GRASP65, a protein involved in the stacking of Golgi cisternae. *Cell* 91: 253-262.
[50] Wang Y, Seemann J, Pypaert M, Shorter J, Warren G (2003) A direct role for GRASP65 as a mitotically regulated Golgi stacking factor. *EMBO J.* 22: 3279-3290.
[51] Xiang Y, Wang Y (2010) GRASP55 and GRASP65 play complementary and essential roles in Golgi cisternal stacking. *J. Cell Biol.* 188: 237-251.
[52] Kondylis V, Spoorendonk KM, Rabouille C (2005) dGRASP localization and function in the early exocytic pathway in Drosophila S2 cells. *Mol. Biol. Cell* 16: 4061-4072.
[53] Wang Y, Satoh A, Warren G (2005) Mapping the functional domains of the Golgi stacking factor GRASP65. *J. Biol. Chem.* 280: 4921-4928.
[54] Tang D, Mar K, Warren G, Wang Y (2008) Molecular mechanism of mitotic Golgi disassembly and reassembly revealed by a defined reconstitution assay. *J. Biol. Chem.* 283: 6085-6094.
[55] Wang Y, Wei JH, Bisel B, Tang D, Seemann J (2008) Golgi cisternal unstacking stimulates COPI vesicle budding and protein transport. *PLoS One* 3: e1647.
[56] Puthenveedu MA, Bachert C, Puri S, Lanni F, Linstedt AD (2006) GM130 and GRASP65-dependent lateral cisternal fusion allows uniform Golgi-enzyme distribution. *Nat. Cell Biol.* 8: 238-248.
[57] Lu Z, Joseph D, Bugnard E, Zaal KJ, Ralston E (2001) Golgi complex reorganization during muscle differentiation: visualization in living cells and mechanism. *Mol. Biol. Cell* 12: 795-808.
[58] Misteli T (2001) The concept of self-organization in cellular architecture. *J. Cell Biol.* 155: 181-185.
[59] Altan-Bonnet N, Sougrat R, Lippincott-Schwartz J (2004) Molecular basis for Golgi maintenance and biogenesis. *Curr. Opin. Cell Biol.* 16: 364-372.
[60] Short B, Haas A, Barr FA (2005) Golgins and GTPases, giving identity and structure to the Golgi apparatus. *Biochim. Biophys. Acta,* 1744: 383-395.
[61] Nakamura N, Rabouille C, Watson R, Nilsson T, Hui N, et al. (1995) Characterization of a cis-Golgi matrix protein, GM130. *J. Cell Biol.* 131: 1715-1726.
[62] Misumi Y, Sohda M, Yano A, Fujiwara T, Ikehara Y (1997) Molecular characterization of GCP170, a 170-kDa protein associated with the cytoplasmic face of the Golgi membrane. *J. Biol. Chem.* 272: 23851-23858.
[63] Fritzler MJ, Hamel JC, Ochs RL, Chan EK (1993) Molecular characterization of two human autoantigens: unique cDNAs encoding 95- and 160-kD proteins of a putative family in the Golgi complex. *J. Exp. Med.* 178: 49-62.

[64] Griffith KJ, Chan EK, Lung CC, Hamel JC, Guo X, et al. (1997) Molecular cloning of a novel 97-kd Golgi complex autoantigen associated with Sjogren's syndrome. *Arthritis. Rheum.* 40: 1693-1702.

[65] Rios RM, Tassin AM, Celati C, Antony C, Boissier MC, et al. (1994) A peripheral protein associated with the cis-Golgi network redistributes in the intermediate compartment upon brefeldin A treatment. *J. Cell Biol.* 125: 997-1013.

[66] Pernet-Gallay K, Antony C, Johannes L, Bornens M, Goud B, et al. (2002) The overexpression of GMAP-210 blocks anterograde and retrograde transport between the ER and the Golgi apparatus. *Traffic,* 3: 822-832.

[67] Erlich R, Gleeson PA, Campbell P, Dietzsch E, Toh BH (1996) Molecular characterization of trans-Golgi p230. A human peripheral membrane protein encoded by a gene on chromosome 6p12-22 contains extensive coiled-coil alpha-helical domains and a granin motif. *J. Biol. Chem.* 271: 8328-8337.

[68] Luke MR, Kjer-Nielsen L, Brown DL, Stow JL, Gleeson PA (2003) GRIP domain-mediated targeting of two new coiled-coil proteins, GCC88 and GCC185, to subcompartments of the trans-Golgi network. *J. Biol. Chem.* 278: 4216-4226.

[69] Short B, Preisinger C, Korner R, Kopajtich R, Byron O, et al. (2001) A GRASP55-rab2 effector complex linking Golgi structure to membrane traffic. *J. Cell Biol.* 155: 877-883.

[70] Jakymiw A, Raharjo E, Rattner JB, Eystathioy T, Chan EK, et al. (2000) Identification and characterization of a novel Golgi protein, golgin-67. *J. Biol. Chem.* 275: 4137-4144.

[71] Bascom RA, Srinivasan S, Nussbaum RL (1999) Identification and characterization of golgin-84, a novel Golgi integral membrane protein with a cytoplasmic coiled-coil domain. *J. Biol. Chem.* 274: 2953-2962.

[72] Linstedt AD, Hauri HP (1993) Giantin, a novel conserved Golgi membrane protein containing a cytoplasmic domain of at least 350 kDa. *Mol. Biol. Cell* 4: 679-693.

[73] Drin G, Morello V, Casella JF, Gounon P, Antonny B (2008) Asymmetric tethering of flat and curved lipid membranes by a golgin. *Science* 320: 670-673.

[74] Moyer BD, Allan BB, Balch WE (2001) Rab1 interaction with a GM130 effector complex regulates COPII vesicle cis--Golgi tethering. *Traffic,* 2: 268-276.

[75] Shorter J, Beard MB, Seemann J, Dirac-Svejstrup AB, Warren G (2002) Sequential tethering of Golgins and catalysis of SNAREpin assembly by the vesicle-tethering protein p115. *J. Cell Biol.* 157: 45-62.

[76] Lu L, Tai G, Hong W (2004) Autoantigen Golgin-97, an effector of Arl1 GTPase, participates in traffic from the endosome to the trans-golgi network. *Mol. Biol. Cell* 15: 4426-4443.

[77] Diao A, Rahman D, Pappin DJ, Lucocq J, Lowe M (2003) The coiled-coil membrane protein golgin-84 is a novel rab effector required for Golgi ribbon formation. *J. Cell Biol.* 160: 201-212.

[78] Yadav S, Puri S, Linstedt AD (2009) A primary role for Golgi positioning in directed secretion, cell polarity, and wound healing. *Mol. Biol. Cell,* 20: 1728-1736.

[79] Hicks SW, Horn TA, McCaffery JM, Zuckerman DM, Machamer CE (2006) Golgin-160 promotes cell surface expression of the beta-1 adrenergic receptor. *Traffic,* 7: 1666-1677.

[80] Williams D, Hicks SW, Machamer CE, Pessin JE (2006) Golgin-160 is required for the Golgi membrane sorting of the insulin-responsive glucose transporter GLUT4 in adipocytes. *Mol. Biol. Cell,* 17: 5346-5355.

[81] Hicks SW, Machamer CE (2005) Isoform-specific interaction of golgin-160 with the Golgi-associated protein PIST. *J. Biol. Chem.* 280: 28944-28951.

[82] Gillingham AK, Munro S (2003) Long coiled-coil proteins and membrane traffic. *Biochim. Biophys. Acta,* 1641: 71-85.

[83] Kreis TE (1990) Role of microtubules in the organisation of the Golgi apparatus. *Cell Motil. Cytoskeleton.* 15: 67-70.

[84] Cole NB, Lippincott-Schwartz J (1995) Organization of organelles and membrane traffic by microtubules. *Curr. Opin. Cell Biol.* 7: 55-64.

[85] Stow JL, Fath KR, Burgess DR (1998) Budding roles for myosin II on the Golgi. *Trends Cell Biol.* 8: 138-141.

[86] Bryant DM, Stow JL (2004) The ins and outs of E-cadherin trafficking. *Trends Cell Biol.* 14: 427-434.

[87] Rogalski AA, Singer SJ (1984) Associations of elements of the Golgi apparatus with microtubules. *J. Cell Biol.* 99: 1092-1100.

[88] Thyberg J, Moskalewski S (1985) Microtubules and the organization of the Golgi complex. *Exp. Cell Res.* 159: 1-16.

[89] Cole NB, Sciaky N, Marotta A, Song J, Lippincott-Schwartz J (1996) Golgi dispersal during microtubule disruption: regeneration of Golgi stacks at peripheral endoplasmic reticulum exit sites. *Mol. Biol. Cell,* 7: 631-650.

[90] Corthesy-Theulaz I, Pauloin A, Pfeffer SR (1992) Cytoplasmic dynein participates in the centrosomal localization of the Golgi complex. *J. Cell Biol.* 118: 1333-1345.

[91] Burkhardt JK, Echeverri CJ, Nilsson T, Vallee RB (1997) Overexpression of the dynamitin (p50) subunit of the dynactin complex disrupts dynein-dependent maintenance of membrane organelle distribution. *J. Cell Biol.* 139: 469-484.

[92] Lippincott-Schwartz J, Cole NB (1995) Roles for microtubules and kinesin in membrane traffic between the endoplasmic reticulum and the Golgi complex. *Biochem. Soc. Trans,* 23: 544-548.

[93] Stauber T, Simpson JC, Pepperkok R, Vernos I (2006) A role for kinesin-2 in COPI-dependent recycling between the ER and the Golgi complex. *Curr. Biol.* 16: 2245-2251.

[94] Hehnly H, Stamnes M (2007) Regulating cytoskeleton-based vesicle motility. *FEBS Lett.* 581: 2112-2118.

[95] Linstedt AD (2004) Positioning the Golgi apparatus. *Cell* 118: 271-272.

[96] Chabin-Brion K, Marceiller J, Perez F, Settegrana C, Drechou A, et al. (2001) The Golgi complex is a microtubule-organizing organelle. *Mol. Biol. Cell,* 12: 2047-2060.

[97] Efimov A, Kharitonov A, Efimova N, Loncarek J, Miller PM, et al. (2007) Asymmetric CLASP-dependent nucleation of noncentrosomal microtubules at the trans-Golgi network. *Dev. Cell,* 12: 917-930.

[98] Suter DM, Schaefer AW, Forscher P (2004) Microtubule dynamics are necessary for SRC family kinase-dependent growth cone steering. *Curr. Biol.* 14: 1194-1199.

[99] Miller PM, Folkmann AW, Maia AR, Efimova N, Efimov A, et al. (2009) Golgi-derived CLASP-dependent microtubules control Golgi organization and polarized trafficking in motile cells. *Nat. Cell Biol.* 11: 1069-1080.

[100] Drabek K, van Ham M, Stepanova T, Draegestein K, van Horssen R, et al. (2006) Role of CLASP2 in microtubule stabilization and the regulation of persistent motility. *Curr. Biol.* 16: 2259-2264.

[101] Bacallao R, Antony C, Dotti C, Karsenti E, Stelzer EH, et al. (1989) The subcellular organization of Madin-Darby canine kidney cells during the formation of a polarized epithelium. *J. Cell Biol.* 109: 2817-2832.

[102] Musch A (2004) Microtubule organization and function in epithelial cells. *Traffic*, 5: 1-9.

[103] Wei JH, Seemann J (2009) Induction of asymmetrical cell division to analyze spindle-dependent organelle partitioning using correlative microscopy techniques. *Nat. Protoc.* 4: 1653-1662.

[104] Shima DT, Cabrera-Poch N, Pepperkok R, Warren G (1998) An ordered inheritance strategy for the Golgi apparatus: visualization of mitotic disassembly reveals a role for the mitotic spindle. *J. Cell Biol.* 141: 955-966.

[105] Jesch SA, Mehta AJ, Velliste M, Murphy RF, Linstedt AD (2001) Mitotic Golgi is in a dynamic equilibrium between clustered and free vesicles independent of the ER. *Traffic*, 2: 873-884.

[106] Percival JM, Hughes JA, Brown DL, Schevzov G, Heimann K, et al. (2004) Targeting of a tropomyosin isoform to short microfilaments associated with the Golgi complex. *Mol. Biol. Cell,* 15: 268-280.

[107] Heimann K, Percival JM, Weinberger R, Gunning P, Stow JL (1999) Specific isoforms of actin-binding proteins on distinct populations of Golgi-derived vesicles. *J. Biol. Chem.* 274: 10743-10750.

[108] Sandoz D, Chailley B, Boisvieux-Ulrich E, Lemullois M, Laine MC, et al. (1988) Organization and functions of cytoskeleton in metazoan ciliated cells. *Biol. Cell,* 63: 183-193.

[109] Stow JL, Heimann K (1998) Vesicle budding on Golgi membranes: regulation by G proteins and myosin motors. *Biochim. Biophys. Acta,* 1404: 161-171.

[110] Achler C, Filmer D, Merte C, Drenckhahn D (1989) Role of microtubules in polarized delivery of apical membrane proteins to the brush border of the intestinal epithelium. *J. Cell Biol.* 109: 179-189.

[111] Fath KR, Burgess DR (1993) Golgi-derived vesicles from developing epithelial cells bind actin filaments and possess myosin-I as a cytoplasmically oriented peripheral membrane protein. *J. Cell Biol.* 120: 117-127.

[112] di Campli A, Valderrama F, Babia T, De Matteis MA, Luini A, et al. (1999) Morphological changes in the Golgi complex correlate with actin cytoskeleton rearrangements. *Cell Motil. Cytoskeleton.* 43: 334-348.

[113] Valderrama F, Babia T, Ayala I, Kok JW, Renau-Piqueras J, et al. (1998) Actin microfilaments are essential for the cytological positioning and morphology of the Golgi complex. *Eur. J. Cell Biol.* 76: 9-17.

[114] Holleran EA, Tokito MK, Karki S, Holzbaur EL (1996) Centractin (ARP1) associates with spectrin revealing a potential mechanism to link dynactin to intracellular organelles. *J. Cell Biol.* 135: 1815-1829.

[115] Beck KA, Nelson WJ (1998) A spectrin membrane skeleton of the Golgi complex. *Biochim. Biophys. Acta,* 1404: 153-160.

[116] De Matteis MA, Morrow JS (1998) The role of ankyrin and spectrin in membrane transport and domain formation. *Curr. Opin. Cell Biol.* 10: 542-549.
[117] Gough LL, Fan J, Chu S, Winnick S, Beck KA (2003) Golgi localization of Syne-1. *Mol. Biol. Cell,* 14: 2410-2424.
[118] De Matteis MA, Morrow JS (2000) Spectrin tethers and mesh in the biosynthetic pathway. *J. Cell Sci.* 113 (Pt 13): 2331-2343.
[119] Gough LL, Beck KA (2004) The spectrin family member Syne-1 functions in retrograde transport from Golgi to ER. *Biochim. Biophys. Acta,* 1693: 29-36.
[120] von Blume J, Duran JM, Forlanelli E, Alleaume AM, Egorov M, et al. (2009) Actin remodeling by ADF/cofilin is required for cargo sorting at the trans-Golgi network. *J. Cell Biol,* 187: 1055-1069.
[121] Salvarezza SB, Deborde S, Schreiner R, Campagne F, Kessels MM, et al. (2009) LIM kinase 1 and cofilin regulate actin filament population required for dynamin-dependent apical carrier fission from the trans-Golgi network. *Mol. Biol. Cell,* 20: 438-451.
[122] Valderrama F, Luna A, Babia T, Martinez-Menarguez JA, Ballesta J, et al. (2000) The golgi-associated COPI-coated buds and vesicles contain beta/gamma -actin. *Proc. Natl. Acad. Sci. U S A,* 97: 1560-1565.
[123] Kondylis V, van Nispen tot Pannerden HE, Herpers B, Friggi-Grelin F, Rabouille C (2007) The golgi comprises a paired stack that is separated at G2 by modulation of the actin cytoskeleton through Abi and Scar/WAVE. *Dev. Cell,* 12: 901-915.
[124] Mulholland J, Wesp A, Riezman H, Botstein D (1997) Yeast actin cytoskeleton mutants accumulate a new class of Golgi-derived secretary vesicle. *Mol. Biol. Cell,* 8: 1481-1499.
[125] Ullman MD, Radin NS (1974) The enzymatic formation of sphingomyelin from ceramide and lecithin in mouse liver. *J. Biol. Chem.* 249: 1506-1512.
[126] Voelker DR, Kennedy EP (1982) Cellular and enzymic synthesis of sphingomyelin. *Biochemistry,* 21: 2753-2759.
[127] Billah MM, Anthes JC (1990) The regulation and cellular functions of phosphatidylcholine hydrolysis. *Biochem. J.* 269: 281-291.
[128] Exton JH (1990) Signaling through phosphatidylcholine breakdown. *J. Biol. Chem.* 265: 1-4.
[129] Liscovitch M (1992) Crosstalk among multiple signal-activated phospholipases. *Trends Biochem. Sci.* 17: 393-399.
[130] Balboa MA, Balsinde J, Dennis EA, Insel PA (1995) A phospholipase D-mediated pathway for generating diacylglycerol in nuclei from Madin-Darby canine kidney cells. *J. Biol. Chem.* 270: 11738-11740.
[131] Baron CL, Malhotra V (2002) Role of diacylglycerol in PKD recruitment to the TGN and protein transport to the plasma membrane. *Science,* 295: 325-328.
[132] Maeda Y, Beznoussenko GV, Van Lint J, Mironov AA, Malhotra V (2001) Recruitment of protein kinase D to the trans-Golgi network via the first cysteine-rich domain. *EMBO J.* 20: 5982-5990.
[133] Bankaitis VA (2002) Cell biology. Slick recruitment to the Golgi. *Science,* 295: 290-291.
[134] Cockcroft S, Taylor JA, Judah JD (1985) Subcellular localisation of inositol lipid kinases in rat liver. *Biochim. Biophys. Acta,* 845: 163-170.

[135] Wang YJ, Wang J, Sun HQ, Martinez M, Sun YX, et al. (2003) Phosphatidylinositol 4 phosphate regulates targeting of clathrin adaptor AP-1 complexes to the Golgi. *Cell* 114: 299-310.

[136] Wang J, Sun HQ, Macia E, Kirchhausen T, Watson H, et al. (2007) PI4P promotes the recruitment of the GGA adaptor proteins to the trans-Golgi network and regulates their recognition of the ubiquitin sorting signal. *Mol. Biol. Cell.* 18: 2646-2655.

[137] Audhya A, Foti M, Emr SD (2000) Distinct roles for the yeast phosphatidylinositol 4-kinases, Stt4p and Pik1p, in secretion, cell growth, and organelle membrane dynamics. *Mol. Biol. Cell,* 11: 2673-2689.

[138] Schorr M, Then A, Tahirovic S, Hug N, Mayinger P (2001) The phosphoinositide phosphatase Sac1p controls trafficking of the yeast Chs3p chitin synthase. *Curr. Biol.* 11: 1421-1426.

[139] Mayinger P (2009) Regulation of Golgi function via phosphoinositide lipids. *Semin. Cell Dev. Biol.* 20: 793-800.

[140] Foti M, Audhya A, Emr SD (2001) Sac1 lipid phosphatase and Stt4 phosphatidylinositol 4-kinase regulate a pool of phosphatidylinositol 4-phosphate that functions in the control of the actin cytoskeleton and vacuole morphology. *Mol. Biol. Cell,* 12: 2396-2411.

[141] Dippold HC, Ng MM, Farber-Katz SE, Lee SK, Kerr ML, et al. (2009) GOLPH3 bridges phosphatidylinositol-4- phosphate and actomyosin to stretch and shape the Golgi to promote budding. *Cell* 139: 337-351.

[142] De Matteis MA, Di Campli A, Godi A (2005) The role of the phosphoinositides at the Golgi complex. *Biochim. Biophys. Acta,* 1744: 396-405.

[143] Toth B, Balla A, Ma H, Knight ZA, Shokat KM, et al. (2006) Phosphatidylinositol 4-kinase IIIbeta regulates the transport of ceramide between the endoplasmic reticulum and Golgi. *J. Biol. Chem.* 281: 36369-36377.

[144] Watt SA, Kular G, Fleming IN, Downes CP, Lucocq JM (2002) Subcellular localization of phosphatidylinositol 4,5-bisphosphate using the pleckstrin homology domain of phospholipase C delta1. *Biochem. J.* 363: 657-666.

[145] Guerriero CJ, Weixel KM, Bruns JR, Weisz OA (2006) Phosphatidylinositol 5-kinase stimulates apical biosynthetic delivery via an Arp2/3-dependent mechanism. *J. Biol. Chem.* 281: 15376-15384.

[146] Godi A, Santone I, Pertile P, Devarajan P, Stabach PR, et al. (1998) ADP ribosylation factor regulates spectrin binding to the Golgi complex. *Proc. Natl. Acad. Sci. U S A,* 95: 8607-8612.

[147] Sweeney DA, Siddhanta A, Shields D (2002) Fragmentation and re-assembly of the Golgi apparatus in vitro. A requirement for phosphatidic acid and phosphatidylinositol 4,5-bisphosphate synthesis. *J. Biol. Chem.* 277: 3030-3039.

[148] Siddhanta A, Radulescu A, Stankewich MC, Morrow JS, Shields D (2003) Fragmentation of the Golgi apparatus. A role for beta III spectrin and synthesis of phosphatidylinositol 4,5-bisphosphate. *J. Biol. Chem.* 278: 1957-1965.

[149] De Matteis M, Godi A, Corda D (2002) Phosphoinositides and the golgi complex. *Curr. Opin. Cell Biol.* 14: 434-447.

[150] Godi A, Pertile P, Meyers R, Marra P, Di Tullio G, et al. (1999) ARF mediates recruitment of PtdIns-4-OH kinase-beta and stimulates synthesis of PtdIns(4,5)P2 on the Golgi complex. *Nat. Cell Biol.* 1: 280-287.

[151] Futerman AH, Pagano RE (1991) Determination of the intracellular sites and topology of glucosylceramide synthesis in rat liver. *Biochem. J.* 280 (Pt 2): 295-302.

[152] Halter D, Neumann S, van Dijk SM, Wolthoorn J, de Maziere AM, et al. (2007) Pre- and post-Golgi translocation of glucosylceramide in glycosphingolipid synthesis. *J. Cell Biol.* 179: 101-115.

[153] Collins RN, Warren G (1992) Sphingolipid transport in mitotic HeLa cells. *J. Biol. Chem.* 267: 24906-24911.

[154] Kok JW, Babia T, Klappe K, Egea G, Hoekstra D (1998) Ceramide transport from endoplasmic reticulum to Golgi apparatus is not vesicle-mediated. *Biochem. J.* 333 (Pt 3): 779-786.

[155] Futerman AH (2006) Intracellular trafficking of sphingolipids: relationship to biosynthesis. *Biochim. Biophys. Acta,* 1758: 1885-1892.

[156] Hanada K, Kumagai K, Yasuda S, Miura Y, Kawano M, et al. (2003) Molecular machinery for non-vesicular trafficking of ceramide. *Nature,* 426: 803-809.

[157] Hanada K, Kumagai K, Tomishige N, Kawano M (2007) CERT and intracellular trafficking of ceramide. *Biochim. Biophys. Acta,* 1771: 644-653.

[158] Munro S (2003) Cell biology: earthworms and lipid couriers. *Nature* 426: 775-776.

[159] Holthuis JC, Levine TP (2005) Lipid traffic: floppy drives and a superhighway. *Nat. Rev. Mol. Cell Biol.* 6: 209-220.

[160] Chandran S, Machamer CE (2008) Acute perturbations in Golgi organization impact de novo sphingomyelin synthesis. *Traffic,* 9: 1894-1904.

[161] Loewen CJ, Roy A, Levine TP (2003) A conserved ER targeting motif in three families of lipid binding proteins and in Opi1p binds VAP. *EMBO J.* 22: 2025-2035.

[162] Lapierre LA, Tuma PL, Navarre J, Goldenring JR, Anderson JM (1999) VAP-33 localizes to both an intracellular vesicle population and with occludin at the tight junction. *J. Cell Sci.* 112 (Pt 21): 3723-3732.

[163] Nishimura Y, Hayashi M, Inada H, Tanaka T (1999) Molecular cloning and characterization of mammalian homologues of vesicle-associated membrane protein-associated (VAMP-associated) proteins. *Biochem. Biophys. Res. Commun.* 254: 21-26.

[164] Kawano M, Kumagai K, Nishijima M, Hanada K (2006) Efficient trafficking of ceramide from the endoplasmic reticulum to the Golgi apparatus requires a VAMP-associated protein-interacting FFAT motif of CERT. *J. Biol. Chem.* 281: 30279-30288.

[165] Perry RJ, Ridgway ND (2006) Oxysterol-binding protein and vesicle-associated membrane protein-associated protein are required for sterol-dependent activation of the ceramide transport protein. *Mol. Biol. Cell,* 17: 2604-2616.

[166] Weir ML, Klip A, Trimble WS (1998) Identification of a human homologue of the vesicle-associated membrane protein (VAMP)-associated protein of 33 kDa (VAP-33): a broadly expressed protein that binds to VAMP. *Biochem. J.* 333 (Pt 2): 247-251.

[167] Skehel PA, Fabian-Fine R, Kandel ER (2000) Mouse VAP33 is associated with the endoplasmic reticulum and microtubules. *Proc. Natl. Acad. Sci. U S A* 97: 1101-1106.

[168] Kumagai K, Kawano M, Shinkai-Ouchi F, Nishijima M, Hanada K (2007) Interorganelle trafficking of ceramide is regulated by phosphorylation-dependent cooperativity between the PH and START domains of CERT. *J. Biol. Chem.* 282: 17758-17766.

[169] Saito S, Matsui H, Kawano M, Kumagai K, Tomishige N, et al. (2008) Protein phosphatase 2Cepsilon is an endoplasmic reticulum integral membrane protein that

dephosphorylates the ceramide transport protein CERT to enhance its association with organelle membranes. *J. Biol. Chem. 283*: 6584-6593.

[170] Fugmann T, Hausser A, Schoffler P, Schmid S, Pfizenmaier K, et al. (2007) Regulation of secretory transport by protein kinase D-mediated phosphorylation of the ceramide transfer protein. *J. Cell Biol.* 178: 15-22.

[171] Yamaji T, Kumagai K, Tomishige N, Hanada K (2008) Two sphingolipid transfer proteins, CERT and FAPP2: their roles in sphingolipid metabolism. *IUBMB Life*, 60: 511-518.

[172] Li MG, Katsura K, Nomiyama H, Komaki K, Ninomiya-Tsuji J, et al. (2003) Regulation of the interleukin-1-induced signaling pathways by a novel member of the protein phosphatase 2C family (PP2Cepsilon). *J. Biol. Chem.* 278: 12013-12021.

[173] Lannert H, Gorgas K, Meissner I, Wieland FT, Jeckel D (1998) Functional organization of the Golgi apparatus in glycosphingolipid biosynthesis. Lactosylceramide and subsequent glycosphingolipids are formed in the lumen of the late Golgi. *J. Biol. Chem.* 273: 2939-2946.

[174] Dowler S, Currie RA, Campbell DG, Deak M, Kular G, et al. (2000) Identification of pleckstrin-homology-domain-containing proteins with novel phosphoinositide-binding specificities. *Biochem. J.* 351: 19-31.

[175] Giussani P, Colleoni T, Brioschi L, Bassi R, Hanada K, et al. (2008) Ceramide traffic in C6 glioma cells: evidence for CERT-dependent and independent transport from ER to the Golgi apparatus. *Biochim. Biophys. Acta,* 1781: 40-51.

[176] Godi A, Di Campli A, Konstantakopoulos A, Di Tullio G, Alessi DR, et al. (2004) FAPPs control Golgi-to-cell-surface membrane traffic by binding to ARF and PtdIns(4)P. *Nat. Cell Biol.* 6: 393-404.

[177] D'Angelo G, Polishchuk E, Di Tullio G, Santoro M, Di Campli A, et al. (2007) Glycosphingolipid synthesis requires FAPP2 transfer of glucosylceramide. *Nature* 449: 62-67.

[178] Goldstein JL, DeBose-Boyd RA, Brown MS (2006) Protein sensors for membrane sterols. *Cell* 124: 35-46.

[179] Perry RJ, Ridgway ND (2005) Molecular mechanisms and regulation of ceramide transport. *Biochim. Biophys. Acta,* 1734: 220-234.

[180] Schroepfer GJ, Jr. (2000) Oxysterols: modulators of cholesterol metabolism and other processes. *Physiol. Rev.* 80: 361-554.

[181] Maxfield FR, Tabas I (2005) Role of cholesterol and lipid organization in disease. *Nature*, 438: 612-621.

[182] Ridgway ND, Dawson PA, Ho YK, Brown MS, Goldstein JL (1992) Translocation of oxysterol binding protein to Golgi apparatus triggered by ligand binding. *J. Cell Biol.* 116: 307-319.

[183] Storey MK, Byers DM, Cook HW, Ridgway ND (1998) Cholesterol regulates oxysterol binding protein (OSBP) phosphorylation and Golgi localization in Chinese hamster ovary cells: correlation with stimulation of sphingomyelin synthesis by 25-hydroxycholesterol. *Biochem. J.* 336 (Pt 1): 247-256.

[184] Nhek S, Ngo M, Yang X, Ng MM, Field SJ, et al. (2010) Regulation of OSBP Golgi Localization through Protein Kinase D-mediated Phosphorylation. *Mol. Biol. Cell.*

[185] Wyles JP, McMaster CR, Ridgway ND (2002) Vesicle-associated membrane protein-associated protein-A (VAP-A) interacts with the oxysterol-binding protein to modify export from the endoplasmic reticulum. *J. Biol. Chem.* 277: 29908-29918.

[186] Rao RP, Yuan C, Allegood JC, Rawat SS, Edwards MB, et al. (2007) Ceramide transfer protein function is essential for normal oxidative stress response and lifespan. *Proc. Natl. Acad. Sci. U S A,* 104: 11364-11369.

[187] Rietveld A, Neutz S, Simons K, Eaton S (1999) Association of sterol- and glycosylphosphatidylinositol-linked proteins with Drosophila raft lipid microdomains. *J. Biol. Chem.* 274: 12049-12054.

[188] Huitema K, van den Dikkenberg J, Brouwers JF, Holthuis JC (2004) Identification of a family of animal sphingomyelin synthases. *EMBO J.* 23: 33-44.

[189] Vacaru AM, Tafesse FG, Ternes P, Kondylis V, Hermansson M, et al. (2009) Sphingomyelin synthase-related protein SMSr controls ceramide homeostasis in the ER. *J. Cell Biol.* 185: 1013-1027.

[190] Wang X, Rao RP, Kosakowska-Cholody T, Masood MA, Southon E, et al. (2009) Mitochondrial degeneration and not apoptosis is the primary cause of embryonic lethality in ceramide transfer protein mutant mice. *J. Cell. Biol.* 184: 143-158.

[191] Marchesini N, Osta W, Bielawski J, Luberto C, Obeid LM, et al. (2004) Role for mammalian neutral sphingomyelinase 2 in confluence-induced growth arrest of MCF7 cells. *J. Biol. Chem.* 279: 25101-25111.

[192] Ferri KF, Kroemer G (2001) Organelle-specific initiation of cell death pathways. *Nat. Cell Biol.* 3: E255-263.

[193] Thornberry NA, Lazebnik Y (1998) Caspases: enemies within. *Science* 281: 1312-1316.

[194] Boatright KM, Salvesen GS (2003) Mechanisms of caspase activation. *Curr. Opin. Cell Biol.* 15: 725-731.

[195] Creagh EM, Conroy H, Martin SJ (2003) Caspase-activation pathways in apoptosis and immunity. *Immunol. Rev.* 193: 10-21.

[196] Sriskandan S, Altmann DM (2008) The immunology of sepsis. *J. Pathol.* 214: 211-223.

[197] Bouchier-Hayes L, Oberst A, McStay GP, Connell S, Tait SW, et al. (2009) Characterization of cytoplasmic caspase-2 activation by induced proximity. *Mol. Cell,* 35: 830-840.

[198] Tinel A, Tschopp J (2004) The PIDDosome, a protein complex implicated in activation of caspase-2 in response to genotoxic stress. *Science* 304: 843-846.

[199] Cory S, Adams JM (2002) The Bcl2 family: regulators of the cellular life-or-death switch. *Nat. Rev. Cancer,* 2: 647-656.

[200] Youle RJ, Strasser A (2008) The BCL-2 protein family: opposing activities that mediate cell death. *Nat. Rev. Mol. Cell Biol.* 9: 47-59.

[201] Sharpe JC, Arnoult D, Youle RJ (2004) Control of mitochondrial permeability by Bcl-2 family members. *Biochim. Biophys. Acta,* 1644: 107-113.

[202] Kuwana T, Newmeyer DD (2003) Bcl-2-family proteins and the role of mitochondria in apoptosis. *Curr. Opin. Cell Biol.* 15: 691-699.

[203] Morishima N, Nakanishi K, Tsuchiya K, Shibata T, Seiwa E (2004) Translocation of Bim to the endoplasmic reticulum (ER) mediates ER stress signaling for activation of caspase-12 during ER stress-induced apoptosis. *J. Biol. Chem.* 279: 50375-50381.

[204] Breckenridge DG, Stojanovic M, Marcellus RC, Shore GC (2003) Caspase cleavage product of BAP31 induces mitochondrial fission through endoplasmic reticulum

calcium signals, enhancing cytochrome c release to the cytosol. *J. Cell Biol.* 160: 1115-1127.
[205] Verhagen AM, Coulson EJ, Vaux DL (2001) Inhibitor of apoptosis proteins and their relatives: IAPs and other BIRPs. *Genome. Biol.* 2: REVIEWS3009.
[206] Salvesen GS (2002) Caspases and apoptosis. Essays Biochem 38: 9-19.
[207] Huang Y, Park YC, Rich RL, Segal D, Myszka DG, et al. (2001) Structural basis of caspase inhibition by XIAP: differential roles of the linker versus the BIR domain. *Cell* 104: 781-790.
[208] Hu S, Yang X (2003) Cellular inhibitor of apoptosis 1 and 2 are ubiquitin ligases for the apoptosis inducer Smac/DIABLO. *J. Biol. Chem.* 278: 10055-10060.
[209] Kumar S, Kinoshita M, Noda M, Copeland NG, Jenkins NA (1994) Induction of apoptosis by the mouse Nedd2 gene, which encodes a protein similar to the product of the Caenorhabditis elegans cell death gene ced-3 and the mammalian IL-1 beta-converting enzyme. *Genes Dev,* 8: 1613-1626.
[210] Lamkanfi M, Declercq W, Kalai M, Saelens X, Vandenabeele P (2002) Alice in caspase land. A phylogenetic analysis of caspases from worm to man. *Cell Death Differ.* 9: 358-361.
[211] Mancini M, Machamer CE, Roy S, Nicholson DW, Thornberry NA, et al. (2000) Caspase-2 is localized at the Golgi complex and cleaves golgin-160 during apoptosis. *J. Cell Biol.* 149: 603-612.
[212] Baliga BC, Colussi PA, Read SH, Dias MM, Jans DA, et al. (2003) Role of prodomain in importin-mediated nuclear localization and activation of caspase-2. *J. Biol. Chem.* 278: 4899-4905.
[213] Paroni G, Henderson C, Schneider C, Brancolini C (2002) Caspase-2 can trigger cytochrome C release and apoptosis from the nucleus. *J. Biol. Chem.* 277: 15147-15161.
[214] Shi M, Vivian CJ, Lee KJ, Ge C, Morotomi-Yano K, et al. (2009) DNA-PKcs-PIDDosome: a nuclear caspase-2-activating complex with role in G2/M checkpoint maintenance. *Cell* 136: 508-520.
[215] Machamer CE (2003) Golgi disassembly in apoptosis: cause or effect? *Trends Cell Biol.* 13: 279-281.
[216] Lowe M, Lane JD, Woodman PG, Allan VJ (2004) Caspase-mediated cleavage of syntaxin 5 and giantin accompanies inhibition of secretory traffic during apoptosis. *J. Cell Sci.* 117: 1139-1150.
[217] Maag RS, Mancini M, Rosen A, Machamer CE (2005) Caspase-resistant Golgin-160 disrupts apoptosis induced by secretory pathway stress and ligation of death receptors. *Mol. Biol. Cell,* 16: 3019-3027.
[218] Hauser HP, Bardroff M, Pyrowolakis G, Jentsch S (1998) A giant ubiquitin-conjugating enzyme related to IAP apoptosis inhibitors. *J. Cell Biol.* 141: 1415-1422.
[219] Bartke T, Pohl C, Pyrowolakis G, Jentsch S (2004) Dual role of BRUCE as an antiapoptotic IAP and a chimeric E2/E3 ubiquitin ligase. *Mol. Cell,* 14: 801-811.
[220] Hao Y, Sekine K, Kawabata A, Nakamura H, Ishioka T, et al. (2004) Apollon ubiquitinates SMAC and caspase-9, and has an essential cytoprotection function. *Nat. Cell Biol.* 6: 849-860.
[221] Hicks SW, Machamer CE (2005) Golgi structure in stress sensing and apoptosis. *Biochim. Biophys. Acta,* 1744: 406-414.

[222] Jones SJ, Ledgerwood EC, Prins JB, Galbraith J, Johnson DR, et al. (1999) TNF recruits TRADD to the plasma membrane but not the trans-Golgi network, the principal subcellular location of TNF-R1. *J. Immunol.* 162: 1042-1048.

[223] Bradley JR, Thiru S, Pober JS (1995) Disparate localization of 55-kd and 75-kd tumor necrosis factor receptors in human endothelial cells. *Am. J. Pathol.* 146: 27-32.

[224] Pohl C, Jentsch S (2008) Final stages of cytokinesis and midbody ring formation are controlled by BRUCE. *Cell* 132: 832-845.

[225] Sesso A, Fujiwara DT, Jaeger M, Jaeger R, Li TC, et al. (1999) Structural elements common to mitosis and apoptosis. *Tissue Cell,* 31: 357-371.

[226] Lane JD, Lucocq J, Pryde J, Barr FA, Woodman PG, et al. (2002) Caspase-mediated cleavage of the stacking protein GRASP65 is required for Golgi fragmentation during apoptosis. *J. Cell Biol.* 156: 495-509.

[227] Chiu R, Novikov L, Mukherjee S, Shields D (2002) A caspase cleavage fragment of p115 induces fragmentation of the Golgi apparatus and apoptosis. *J. Cell Biol.* 159: 637-648.

[228] Walker A, Ward C, Sheldrake TA, Dransfield I, Rossi AG, et al. (2004) Golgi fragmentation during Fas-mediated apoptosis is associated with the rapid loss of GM130. *Biochem. Biophys. Res. Commun.* 316: 6-11.

[229] Nozawa K, Casiano CA, Hamel JC, Molinaro C, Fritzler MJ, et al. (2002) Fragmentation of Golgi complex and Golgi autoantigens during apoptosis and necrosis. *Arthritis. Res.* 4: R3.

[230] Lane JD, Vergnolle MA, Woodman PG, Allan VJ (2001) Apoptotic cleavage of cytoplasmic dynein intermediate chain and p150(Glued) stops dynein-dependent membrane motility. *J. Cell Biol.* 153: 1415-1426.

[231] Hicks SW, Machamer CE (2002) The NH2-terminal domain of Golgin-160 contains both Golgi and nuclear targeting information. *J. Biol. Chem.* 277: 35833-35839.

[232] Kondylis V, Rabouille C (2003) A novel role for dp115 in the organization of tER sites in Drosophila. *J. Cell Biol.* 162: 185-198.

[233] Adam D, Heinrich M, Kabelitz D, Schutze S (2002) Ceramide: does it matter for T cells? *Trends Immunol.* 23: 1-4.

[234] Spiegel S, Milstien S (2003) Sphingosine-1-phosphate: an enigmatic signalling lipid. *Nat. Rev. Mol. Cell Biol.* 4: 397-407.

[235] Hannun YA, Obeid LM (2008) Principles of bioactive lipid signalling: lessons from sphingolipids. *Nat. Rev. Mol. Cell Biol.* 9: 139-150.

[236] Bartke N, Hannun YA (2009) Bioactive sphingolipids: metabolism and function. *J. Lipid. Res.* 50 Suppl: S91-96.

[237] Tafesse FG, Ternes P, Holthuis JC (2006) The multigenic sphingomyelin synthase family. *J. Biol. Chem,* 281: 29421-29425.

[238] Mao C, Obeid LM (2008) Ceramidases: regulators of cellular responses mediated by ceramide, sphingosine, and sphingosine-1-phosphate. *Biochim. Biophys. Acta,* 1781: 424-434.

[239] Hait NC, Oskeritzian CA, Paugh SW, Milstien S, Spiegel S (2006) Sphingosine kinases, sphingosine 1-phosphate, apoptosis and diseases. *Biochim. Biophys. Acta,* 1758: 2016-2026.

[240] Bandhuvula P, Saba JD (2007) Sphingosine-1-phosphate lyase in immunity and cancer: silencing the siren. *Trends Mol. Med.* 13: 210-217.

[241] Shinghal R, Scheller RH, Bajjalieh SM (1993) Ceramide 1-phosphate phosphatase activity in brain. *J. Neurochem.* 61: 2279-2285.
[242] Marchesini N, Hannun YA (2004) Acid and neutral sphingomyelinases: roles and mechanisms of regulation. *Biochem. Cell Biol.* 82: 27-44.
[243] Tettamanti G (2004) Ganglioside/glycosphingolipid turnover: new concepts. *Glycoconj. J.* 20: 301-317.
[244] Okazaki T, Bielawska A, Bell RM, Hannun YA (1990) Role of ceramide as a lipid mediator of 1 alpha,25-dihydroxyvitamin D3-induced HL-60 cell differentiation. *J. Biol. Chem.* 265: 15823-15831.
[245] Hetz CA, Hunn M, Rojas P, Torres V, Leyton L, et al. (2002) Caspase-dependent initiation of apoptosis and necrosis by the Fas receptor in lymphoid cells: onset of necrosis is associated with delayed ceramide increase. *J. Cell Sci.* 115: 4671-4683.
[246] Obeid LM, Linardic CM, Karolak LA, Hannun YA (1993) Programmed cell death induced by ceramide. *Science* 259: 1769-1771.
[247] Chalfant CE, Kishikawa K, Mumby MC, Kamibayashi C, Bielawska A, et al. (1999) Long chain ceramides activate protein phosphatase-1 and protein phosphatase-2A. Activation is stereospecific and regulated by phosphatidic acid. *J. Biol. Chem.* 274: 20313-20317.
[248] Wang G, Silva J, Krishnamurthy K, Tran E, Condie BG, et al. (2005) Direct binding to ceramide activates protein kinase Czeta before the formation of a pro-apoptotic complex with PAR-4 in differentiating stem cells. *J. Biol. Chem.* 280: 26415-26424.
[249] Blazquez C, Galve-Roperh I, Guzman M (2000) De novo-synthesized ceramide signals apoptosis in astrocytes via extracellular signal-regulated kinase. *FASEB J.* 14: 2315-2322.
[250] Ruvolo PP (2003) Intracellular signal transduction pathways activated by ceramide and its metabolites. *Pharmacol. Res.* 47: 383-392.
[251] Gomez-Munoz A (2006) Ceramide 1-phosphate/ceramide, a switch between life and death. *Biochim. Biophys. Acta,* 1758: 2049-2056.
[252] Chalfant CE, Spiegel S (2005) Sphingosine 1-phosphate and ceramide 1-phosphate: expanding roles in cell signaling. *J. Cell Sci.* 118: 4605-4612.
[253] Ma Y, Pitson S, Hercus T, Murphy J, Lopez A, et al. (2005) Sphingosine activates protein kinase A type II by a novel cAMP-independent mechanism. *J. Biol. Chem.* 280: 26011-26017.
[254] El Alwani M, Wu BX, Obeid LM, Hannun YA (2006) Bioactive sphingolipids in the modulation of the inflammatory response. *Pharmacol. Ther.* 112: 171-183.
[255] Pyne S, Pyne N (2000) Sphingosine 1-phosphate signalling via the endothelial differentiation gene family of G-protein-coupled receptors. *Pharmacol. Ther.* 88: 115-131.
[256] Hughes H, Budnik A, Schmidt K, Palmer KJ, Mantell J, et al. (2009) Organisation of human ER-exit sites: requirements for the localisation of Sec16 to transitional ER. *J. Cell Sci.* 122: 2924-2934.
[257] Pettus BJ, Chalfant CE, Hannun YA (2002) Ceramide in apoptosis: an overview and current perspectives. *Biochim. Biophys. Acta,* 1585: 114-125.
[258] van Blitterswijk WJ, van der Luit AH, Veldman RJ, Verheij M, Borst J (2003) Ceramide: second messenger or modulator of membrane structure and dynamics? *Biochem. J.* 369: 199-211.

[259] Tepper AD, de Vries E, van Blitterswijk WJ, Borst J (1999) Ordering of ceramide formation, caspase activation, and mitochondrial changes during CD95- and DNA damage-induced apoptosis. *J. Clin. Invest,* 103: 971-978.

[260] Zhang J, Alter N, Reed JC, Borner C, Obeid LM, et al. (1996) Bcl-2 interrupts the ceramide-mediated pathway of cell death. *Proc. Natl. Acad. Sci. U S A* 93: 5325-5328.

[261] Schwandner R, Wiegmann K, Bernardo K, Kreder D, Kronke M (1998) TNF receptor death domain-associated proteins TRADD and FADD signal activation of acid sphingomyelinase. *J. Biol. Chem.* 273: 5916-5922.

[262] Andrieu-Abadie N, Gouaze V, Salvayre R, Levade T (2001) Ceramide in apoptosis signaling: relationship with oxidative stress. *Free Radic. Biol. Med* 31: 717-728.

[263] Andrieu-Abadie N, Levade T (2002) Sphingomyelin hydrolysis during apoptosis. *Biochim. Biophys. Acta,* 1585: 126-134.

[264] Adam-Klages S, Adam D, Wiegmann K, Struve S, Kolanus W, et al. (1996) FAN, a novel WD-repeat protein, couples the p55 TNF-receptor to neutral sphingomyelinase. *Cell* 86: 937-947.

[265] Wright SD, Kolesnick RN (1995) Does endotoxin stimulate cells by mimicking ceramide? *Immunol. Today,* 16: 297-302.

[266] Ganesan V, Colombini M (2010) Regulation of ceramide channels by Bcl-2 family proteins. *FEBS Lett.* 584: 2128-2134.

[267] Fukunaga T, Nagahama M, Hatsuzawa K, Tani K, Yamamoto A, et al. (2000) Implication of sphingolipid metabolism in the stability of the Golgi apparatus. *J. Cell Sci.* 113 (Pt 18): 3299-3307.

[268] Hu W, Xu R, Zhang G, Jin J, Szulc ZM, et al. (2005) Golgi fragmentation is associated with ceramide-induced cellular effects. *Mol. Biol. Cell* 16: 1555-1567.

[269] Lafont E, Milhas D, Carpentier S, Garcia V, Jin ZX, et al. (2010) Caspase-mediated inhibition of sphingomyelin synthesis is involved in FasL-triggered cell death. *Cell Death Differ.* 17: 642-654.

[270] Granero F, Revert F, Revert-Ros F, Lainez S, Martinez-Martinez P, et al. (2005) A human-specific TNF-responsive promoter for Goodpasture antigen-binding protein. *FEBS J.* 272: 5291-5305.

[271] Raya A, Revert F, Navarro S, Saus J (1999) Characterization of a novel type of serine/threonine kinase that specifically phosphorylates the human goodpasture antigen. *J. Biol. Chem.* 274: 12642-12649.

[272] Charruyer A, Bell SM, Kawano M, Douangpanya S, Yen TY, et al. (2008) Decreased ceramide transport protein (CERT) function alters sphingomyelin production following UVB irradiation. *J. Biol. Chem.* 283: 16682-16692.

[273] Kodani A, Sutterlin C (2008) The Golgi protein GM130 regulates centrosome morphology and function. *Mol. Biol. Cell* 19: 745-753.

[274] Follit JA, San Agustin JT, Xu F, Jonassen JA, Samtani R, et al. (2008) The Golgin GMAP210/TRIP11 anchors IFT20 to the Golgi complex. *PLoS Genet.* 4: e1000315.

[275] Follit JA, Tuft RA, Fogarty KE, Pazour GJ (2006) The intraflagellar transport protein IFT20 is associated with the Golgi complex and is required for cilia assembly. *Mol. Biol. Cell,* 17: 3781-3792.

[276] Nachury MV, Loktev AV, Zhang Q, Westlake CJ, Peranen J, et al. (2007) A core complex of BBS proteins cooperates with the GTPase Rab8 to promote ciliary membrane biogenesis. *Cell,* 129: 1201-1213.

Chapter 2

FUNCTIONAL RELATIONSHIPS BETWEEN GOLGI DYNAMICS AND LIPID METABOLISM AND TRANSPORT

Asako Goto and Neale Ridgway[*]

The Atlantic Research Center, Departments of Pediatrics and Biochemistry and Molecular Biology, Dalhousie University, Halifax, Nova Scotia, Canada

ABSTRACT

The Golgi apparatus is a sorting nexus for protein and lipids exported from the endoplasmic reticulum (ER) to other organelles and for secretion. The lipids and sterols that delineate the vesicular/tubular transport carriers and cisternae that constitute the Golgi transport apparatus are just packaging materials but participate directly in membrane fusion, cargo sorting and polarized transport. Low abundance lipids, such as diacylglycerol (DAG), phosphatidic acid (PtdOH), lyso-phospholipids and phosphatidylinositol phosphates, contribute to these processes by localized synthesis and interconversion. These lipids alter the structure of membranes by assisting in induction of positive and negative curvature required for carrier assembly, and regulate the activity of proteins that temporally and spatially regulate fusion and fission events. The Golgi apparatus is especially enriched in phosphatidylinositol 4-phosphate (PtdIns(4)P), where localized metabolism by Golgi-associated PtdIns 4-kinases (PI4K) and phosphatases controls PtdIns(4)P pools that recruit proteins involved in lipid transport and vesicular trafficking. DAG and ceramide conversion in the late Golgi and trans-Golgi network (TGN) by sphingomyelin (SM) synthase regulates Golgi trafficking by recruiting and activating protein kinase D (PKD) for phosphorylation of targets such a PI4KIIIβ. SM and glycosphingolipids (GSL) synthesized in the Golgi apparatus condenses with cholesterol into nanoscale assemblies called lipid rafts. These platforms function in membrane signaling and regulate trans-Golgi network (TGN)-sorting machinery. There is an increasing appreciation for the role of lipid and sterol transfer proteins in modulation of Golgi apparatus function. In particular, site-directed ceramide and sterol transfer

---

[*] Corresponding author. The Atlantic Research Center, Departments of Pediatrics and Biochemistry & Molecular Biology, Dalhousie University, Room C306 CRC Building, Halifax, Nova Scotia, Canada B3H 4H7. Fax: +1-902-494-1394, E-mail address: nridgway@dal.ca

proteins that communicate lipid status, and regulate cholesterol, SM and GSL metabolism. Here we will review the highly integrated lipid metabolic and signaling pathways housed in the Golgi apparatus that control secretory activity and membrane assembly.

ABBREVIATIONS

AP: adaptor protein
Arf: ADP-ribosylation factor
ArfGAP: GTPase-activating proteins
BARS: brefeldin A-induced ADP-ribosylation substrate
BIG: brefeldin A-inhibited guanine nucleotide exchange protein
CERT: ceramide transfer protein
COP: coatomer complex
DAG: diacylglycerol
ENTH: epsin N-terminal homology
EpsinR: Epsin-related protein
FAPP: four-phosphate adaptor protein
FFAT: two phenylalanines in an acidic tract
GBF1: guanine nucleotide exchange factor 1
GCS: UDP-Glc:glucosylceramide synthase
GEF: guanine nucleotide exchange factor
GGA: Golgi-localized γ-ear containing Arf-binding proteins
GlcCer: glucosylceramide
GPAT: glycerol-3-phosphate acyltransferase
GSL: glycosphingolipids
LacCer: lactosylceramide
LPAAT: acyl-CoA-dependent lyso-lipid acyltransferase
LPP: PtdOH phosphatase 2 (PAP2)/lipid phosphate phosphatase
Nir2: PYK2 N-terminal domain-interacting receptor
ORP: OSBP-related protein
OSBP: oxysterol binding protein
PAP: PtdOH phosphatase
PI4K: PtdIns 4-kinase
PH: pleckstrin homology
PITP: PI-transfer protein
PKD: protein kinase D
PLD: phospholipase D
PtdCho: phosphatidylcholine
PtdIns: phosphatidylinositol
PtdOH: phosphatidic acid
SM: sphingomyelin
SMS: SM synthase
SNARE: soluble *N*-ethylmaleimide-sensitive fusion protein (NSF) attachment protein receptor
SREBP: sterol response element-binding protein
VAP: vesicle-associated membrane protein-associated protein

1. INTRODUCTION

The Golgi apparatus is a highly dynamic complex of flattened fenestrated cisternae interconnected by tubules and transport intermediates that sort proteins and lipids to cellular organelles and the external environment (Glick and Nakano, 2009). The organization and shape of these membrane compartments minimizes the internal volume and maximizes surface area, which allows for efficient and rapid protein and lipid trafficking in and out of the organelle. Thus, regulation of membrane shape and curvature is important for trafficking. The processes of membrane trafficking (budding, tubulation, tethering, fusion, and fission) are controlled by membrane deforming proteins, and by production and consumption of specific lipids that modulate charge, curvature and fluidity of the membrane bilayer and recruitment of regulatory proteins (Corda *et al.*, 2002). In particular, phosphatidic acid (PtdOH), diacylglycerol, lyso-phospholipids and PtdIns phosphates (PtdInsPs) are considered to play a key role in controlling Golgi-mediated transport. These lipids are generated locally by Golgi specific kinases, phosphatases, lipases, where they modulate the physical properties of membranes and regulate the recruitment and activity of the secretory machinery. Selective sorting of lipids by vesicular and protein-mediated transport ensures their unique enrichment in organelle membranes, including the Golgi cisternae, that is indispensable for maintaining their structural organization and functions (Wang *et al.*, 2000b; Lev, 2006). For example, the mammalian plasma membrane (PM) is rich in cholesterol, SM and GSLs, whereas the ER is relatively cholesterol poor (van Meer *et al.*, 2008). This heterogeneity in lipid distribution is important from the point of view of maintaining membrane function and homeostasis. Condensation of cholesterol, SM, and GSL at the PM forms lipid rafts that recruit enzymes and signaling factors with a propensity to associate with these liquid-ordered domains (Roper *et al.*, 2000). In contrast, maintaining cholesterol at relatively low levels in the ER allows the resident sterol sensing machinery to respond rapidly to sterol influx and regulate *de novo* synthesis and uptake accordingly (Goldstein *et al.*, 2006a)

In this chapter, we will focus on the interaction of proteins and lipids in the Golgi apparatus and the role in organelle structure and cargo flow. We will also mention human genetic disorders related to lipid metabolism and Golgi function.

2. LIPIDS AND PROTEINS INVOLVED IN MEMBRANE BUDDING

Proteins have been identified that physically affect membrane structure or modulate lipid composition to promote sorting functions of the Golgi. Mechanistically this involves 1) protein-mediated stabilization of membrane curvature (Sections 2.1. and 2.2.), 2) asymmetrical distribution of lipids in the inner and outer leaflets of a bilayer (Section 2.3.), and 3) membrane tethering and fusion mediated by proteins (Sections 3.1. and 3.2.). We will describe these mechanisms in the context of membrane trafficking function of the Golgi, however these are general mechanisms that affect organelle shape throughout the cell.

Protein and lipid transport is mediated by three mechanisms: vesicular trafficking, cisternal maturation, and tubular cisternal connections. We will focus on vesicular trafficking in the current chapter, however readers are referred to excellent recent reviews on other mechanisms of Golgi transport (Puthenveedu and Linstedt, 2005; Glick and Nakano, 2009).

Three types of coated vesicle carriers have been characterized in detail with regard to their structural components and functions: adaptor/clathrin-coated vesicles that mediate transport in the late secretory pathway (Golgi to endosome) and the endocytic pathway (endosome to PM) (Robinson, 2004), coatomer complex (COP) I-coated vesicles in the early secretory pathway (transport through Golgi cisternae and ER to/from Golgi trafficking) (Beck et al., 2009), and COPII-coated vesicles that export proteins from the ER to the Golgi apparatus (Hughes and Stephens, 2008) (Fig.1). An important difference between clathrin-coats and COP-coats is that in most cases cells survive without individual adaptors or clathrin, whereas COPI and COPII are essential. Thus, adaptor/clathrin pathways are more versatile than COP systems, which play specialized roles and are highly conserved among various organisms.

Coat proteins are required for cargo selection and membrane deformation to mediate budding of transport vesicles from a donor compartment. An initiating step in this process is the recruitment of COPI and clathrin to Golgi membranes by the Arf1 GTPases, adaptor proteins (APs) and Golgi-localized γ-ear containing ADP-ribosylation factor (Arf1)-binding proteins (GGAs) (Traub et al., 1993; Dascher and Balch, 1994; Dell'Angelica et al., 2000). On the other hand, the GTPase Sar1 is involved in recruitment of COPII to ER exit sites (Pathre et al., 2003). Upon initiation of recruitment, Arf1 is converted from a GDP- to GTP-bound form by a guanine nucleotide exchange factor (GEF) that is recruited to the Golgi membrane in a Rab1-dependent manner. Upon exchange of GDP for GTP, the folded myristoylated N-terminus helix of Arf1 becomes exposed for membrane anchoring (Liu et al., 2009). GTP hydrolysis causes retraction of the myristoylated N-terminus helix and Arf1 dissociates from the membrane. Specificity of recruitment mediated by Arf1 depends on GEFs. AP-1 recruitment is dependent on brefeldin A-inhibited guanine nucleotide exchange protein (BIG) 1 and 2 that reside in the *trans*-Golgi (Shinotsuka et al., 2002), whereas COPI recruitment is dependent on guanine nucleotide exchange factor 1 (GBF1) that reside in the *cis*-Golgi (Lefrancois and McCormick, 2007; Deng et al., 2009). Recruitment of GGAs is affected either by BIGs and GBF1 (Lefrancois and McCormick, 2007; Manolea et al., 2008). The GTPase Rab1 contributes to the specificity and timing of GBF1 recruitment to the Golgi by activating PI4KIIIα (Dumaresq-Doiron et al., 2010), although it is localized on the nuclear envelope and perinuclear ER (Balla and Balla, 2006). PtdIns(4)P serves as a binding site for GBF1 at the Golgi apparatus, and depletion of PI4KIIIα but not PI4KIIIβ abolished GBF1-dependent recruitment of GGA3 to the TGN. It is not known whether PI4KIIIα is also involved in GBF1-dependent recruitment of COPI to the Golgi apparatus. On the other hand, PI4KIIIβ is involved in BIG1/2 dependent recruitment of AP-1 to the TGN. Rab1 is also involved in triggering tethering of transport vesicles to target membranes (Section 3.2.2.1.).

In addition to the coat proteins that cover the surface of membranes as curvature inducing scaffolds, membrane tubulation and budding that accompanies coatomer assembly also involves protein factors that physically deform membranes (Antonny, 2006). Examples are Epsin-related protein (EpsinR) for adaptor/clathrin-coated vesicles (Section 3.2.1.1.), GTPase-activating proteins (ArfGAPs) and brefeldin A-induced ADP-ribosylation substrate (BARS) for COPI-coated vesicles (Section 3.2.1.2.), and Sar1 for COPII-coated vesicles (Bi et al., 2002; Bielli et al., 2005; Lee et al., 2005). EpsinR, ArfGAP1, and Sar1 generate local positive curvature by the bilayer-couple effect. As hydrophobic interactions between the inner and the outer leaflets of a bilayer tend to keep them coupled to each other, an increase in area

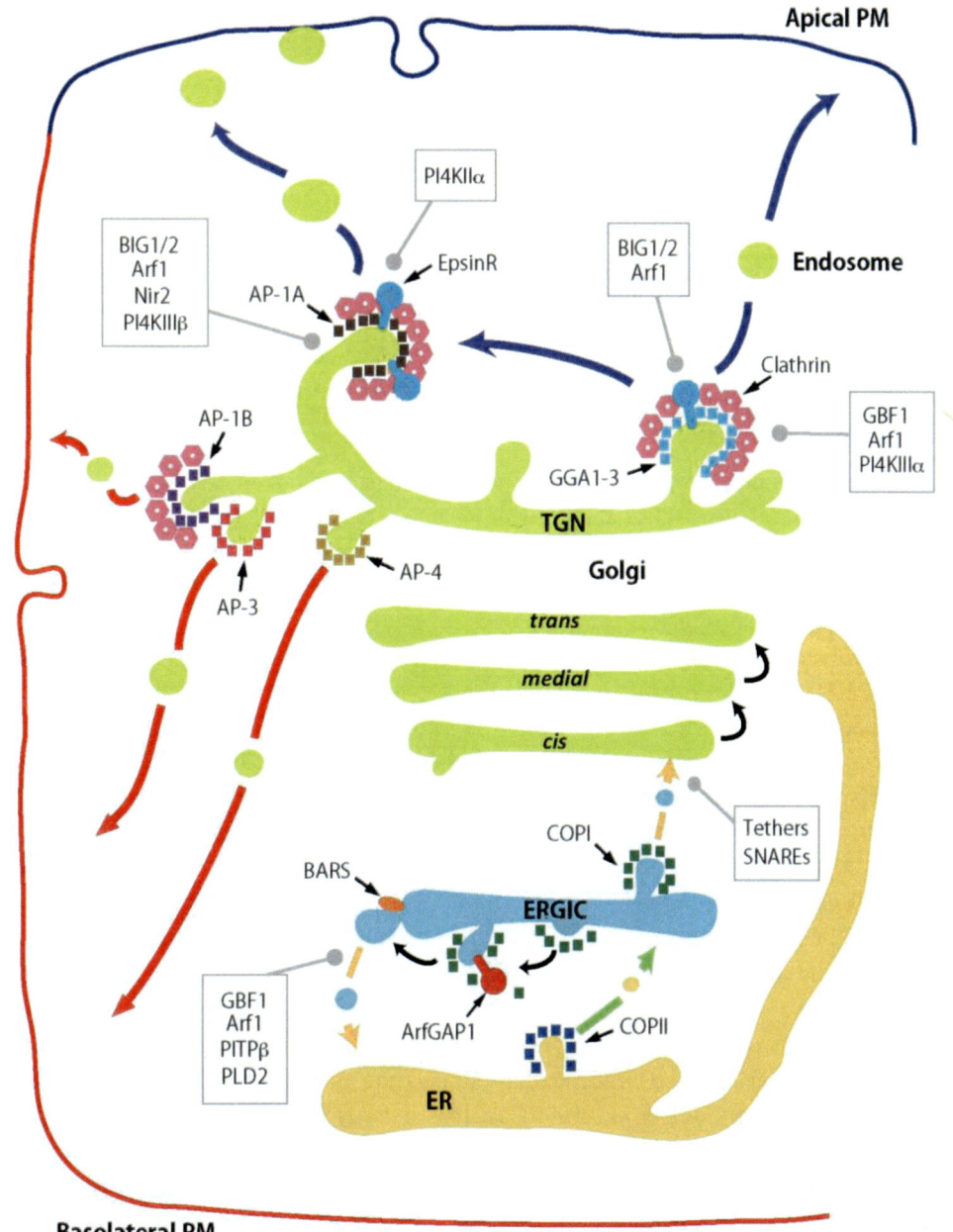

Figure 1. Vesicle trafficking in the Golgi secretory pathway. AP-1A and GGA1-3 cooperate with clathrin to mediate budding of cargo vesicles from the TGN to apical PM. EpsinR is involved in a late stage of AP-1A/clathrin vesicle formation. AP-1B, AP-3 and AP-4 select cargo for transport from the TGN to basolateral PM. AP-1B recruits clathrin for vesicle formation, whereas AP-3 and AP-4 function independently of clathrin. COPI mediates retrograde vesicle trafficking from the *cis*-Golgi to ER. ArfGAP1 and BARS are involved in early and late stages of COPI vesicle formation respectively. COPI is also involved in anterograde vesicle trafficking from ER to *cis*-Golgi in concert with COPII. COPII vesicle transfer cargo from ER to ERGIC and subsequently repackaged in COPI vesicle for transfer form the ERGIC to *cis*-Golgi. Arf1 is responsible for recruitment of adaptor and coat proteins to the site of vesicle formation. BIG1/2 and GBF1 are activators of Arf1 that determine the specificity of protein recruitment. PITPs and PI4Ks are also involved in this process. Refer to the text for details.

by insertion of their amphipathic α-helices into either of the leaflets causes bilayer bending (Zimmerberg and Kozlov, 2006). These membrane-deforming proteins are adapted to membranes of different physio-chemical properties. ER and *cis*-Golgi membranes, where COPI and COPII coats act, are symmetric and the packing between lipids is relatively loose due to low concentration of cholesterol (Bretscher and Munro, 1993). In contrast, *trans*-Golgi/TGN membranes, where adaptor/clathrin-coat acts, is asymmetric, thicker and less permeable due to high concentrations of cholesterol (Bretscher and Munro, 1993). In this chapter, we will focus on adaptor/clathrin-coated vesicles and COPI-coated vesicles that directly contact the Golgi apparatus.

2.1. Adaptor/Clathrin-Coat Vesicle Formation at the Golgi

Clathrin binds indirectly to membrane budding sites through interactions with adaptor proteins. Five APs (AP-1A, GGA1-3 and EpsinR) are regulated by PtdIns(4)P and function in generating adaptor/clathrin-coated vesicles (Fig.1). In addition, most epithelial cells contain AP-1B that differs from AP-1A by having a μ1B instead of μ1A subunit (Folsch *et al.*, 2003). AP-1A is involved in sorting from the TGN to the endosomes and apical surfaces, whereas AP-1B is involved in sorting from the TGN to the basolateral surface (Folsch *et al.*, 2003; Wang *et al.*, 2003) (Fig.1). EpsinR is involved in AP-1A mediated trafficking of cathepsin D from the Golgi to lysosome (Mills *et al.*, 2003). Another role of EpsinR is post-TGN trafficking of soluble *N*-ethylmaleimide-sensitive fusion protein (NSF) attachment protein receptors (SNAREs) (Miller *et al.*, 2007). GGAs are involved in anterograde and retrograde trafficking from TGN to endosomes and vice versa (Wahle *et al.*, 2005; Wang *et al.*, 2007). Adaptor proteins recognize specific signals on cargo proteins at the TGN for recruitment into transport vesicles. APs recognize Yxxϕ (where ϕ is a bulky hydrophobic residue) and D/ExxxLL, and GGAs recognize DxxLL (Robinson, 2004). EpsinR does not recognize cargo motifs but instead interacts with SNAREs by a surface-surface association (Miller *et al.*, 2007).

γ-Adaptin is a component of the AP-1A complex that interacts with clathrin (Wang *et al.*, 2003). Arf1-dependent and Arf1-independent signals are necessary for AP-1A association with Golgi membranes (Godi *et al.*, 1999). In the first Arf1-dependent signal, GTP-bound Arf1 activates PI4KIIIβ and binds AP-1A together with PtdIns(4)P and sorting signals on the selected cargo (Lee *et al.*, 2008). This in turn leads to coat recruitment on membranes followed by membrane remodeling and transport carrier formation. The second signal modulating AP-1A binding to the TGN is dependent on generation of PtdIns(4)P synthesized by PI4KIIα, an Arf1-independent PI4K (Wang *et al.*, 2003). RNAi knockdown of PI4KIIα but not PI4KIIIβ decreased Golgi PtdIns(4)P, and blocked the recruitment of AP-1A complexes to the Golgi. Structural analysis of AP-1 and AP-2 highlighted the importance of the notch created by helix two and three of the γ-subunit for membrane association (Heldwein *et al.*, 2004). Mutation of amino acids that are important in AP-2 counterparts prevented PtdIns(4)P-dependent AP-1A recruitment to the TGN, but did not affect Arf1-dependent binding to liposomes. Two additional adaptor proteins, AP-3 and AP-4 (Robinson, 2004), localize on TGN/endosomal membranes and are involved in sorting to the basolateral surface (Robinson, 2004). Additionally, AP-3 mediates TGN-to-lysosome vesicular trafficking of lysosomal membrane proteins such as lysosome-associated membrane proteins (LAMPs) (Peden et al., 2004). Although they are suggested to function independently of clathrin (Robinson, 2004), the precise mechanism is unknown.

GGA1-3 are monomeric clathrin adaptors that localize in the TGN (Boman et al., 2000; Dell'Angelica et al., 2000; Hirst et al., 2000). As depletion of any one GGA results in a partial decrease in the levels of other GGAs, it has been suggested that GGA1-3 function cooperatively in cargo sorting (Ghosh et al., 2003). GGAs consists of four functional regions; an N-terminus Vps27/Hrs/Starn (VHS) domain that interacts with an acidic cluster DxxLL motif found in the cytoplasmic tail of TGN sorting receptors, a GAT domain that interacts with ubiquitin, PtdIns(4)P and Arf1, a hinge region with clathrin binding motifs, and a C-terminal ear (GAE) domain. The GAE domain has a similar fold as that in γ-adaptin, however the overall structures of GAE and γ-adaptin are very different (Collins et al., 2003). GGAs distribution in the Golgi differs from APs and affects distinct pathways such as membrane trafficking between the TGN and endosomes (Wang et al., 2007), ubiquitin-dependent sorting of cargo proteins both in biosynthetic and endocytic pathways (Kawasaki et al., 2005), and TGN to endosome retrograde trafficking (Wahle et al., 2005). However, GGAs are also adaptor proteins that select cargo for incorporation into AP-1/clathrin-coated vesicles (Doray et al., 2002). Mutant mannose 6-phosphate receptors defective in binding GGAs were poorly incorporated into AP-1/clathrin-coated vesicles, which are responsible for Golgi-to-lysosome sorting.

In addition to coat proteins and their effectors, Epsins are are also involved in the physical deformation of membranes in the TGN to generate the positively curved surfaces of AP/clathrin-coated vesicles (Fig.1). Epsins recruit and promote clathrin polymerization, but less uniformly than APs. EpsinR was originally discovered by a database search and shown to distribute to the Golgi apparatus (Ford et al., 2002; Mills et al., 2003). EpsinR has an Epsin N-terminal homology (ENTH) domain that is shared amongst Epsin family members (Horvath et al., 2007). However, it lacks the C-terminus Eps15 binding motif found in Epsin1-3, instead having a methionine-rich domain with an unknown function (Kent et al., 2002; Nogi et al., 2002; Mills et al., 2003). ENTH domains are found in a number of proteins that induce positive membrane curvature by interacting with PtdInsP. The ENTH domain of Epsin1 binds PtdIns(4,5)P_2 and causes invagination of the PM (Ford et al., 2002). Structural studies showed that an α-helix at the N-terminus of the ENTH domain termed α0 becomes ordered with the other helices upon binding PtdIns(4,5)P_2, and mutations in hydrophobic region in α0 abolished membrane curving ability of Epsin1 (Hyman et al., 2000; Ford et al., 2002). Binding of the ENTH domain to PtdIns(4,5)P_2 allows the α0 region to insert into the outer leaflet of the lipid bilayer, pushing the lipid head groups apart (Ford et al., 2002; Mills et al., 2003). This reduces the energy input necessary to generate highly curved vesicular membranes. The ENTH domain of EpsinR induces curvature at the outer leaflet of the TGN membrane by a similar mechanism as Epsin1. However, unlike other Epsin-family proteins, the PtdIns(4,5)P_2-binding residues within α0 are not conserved in EpsinR. Indeed, EpsinR exhibits a unique preference for PtdIns(4)P that is abundant in the Golgi apparatus (Hirst et al., 2003; Mills et al., 2003). Mutations within the corresponding hydrophobic region in EpsinR α0 cause defective Arf-dependent PtdIns(4)P binding and distribution to the Golgi apparatus (Hirst et al., 2003; Mills et al., 2003). EpsinR binds to AP-1A and GGA2 through its γ-appendage binding motif (D/EFxDF/W) (Mills et al., 2003). As EpsinR knockdown dramatically reduces the amount of AP-1A in clathrin-coated vesicles and vice versa, EpsinR and AP-1A are dependent on each other for maximum incorporation into vesicles formed at the TGN (Hirst et al., 2004), thus promoting membrane curvature and cargo recruitment in a co-operative manner.

2.2. Formation of COPI Vesicles

COPI vesicle formation initiates with assembly of a large heptameric coat complex termed coatomer that is comparable to the organization of the heterotetrameric adaptor/clathrin complex (Eugster et al., 2000). COPI polymerization induces initial membrane deformation (Manneville et al., 2008), and further membrane fission is mediated by activities of ArfGAPs, BARS and PLD2 (Fig.1). Arf1 is a regulating factor that interacts with and recruits the COPI complex to the Golgi, and ArfGAPs regulate the Arf1 GTPase cycle. ArfGAP1-3 catalyze Arf1-bound GTP hydrolysis that triggers uncoating of vesicles prior to fusion (Cukierman et al., 1995; Liu et al., 2001; Watson et al., 2004). ArfGAP1 is recruited to the Golgi cisternae through an amphipathic α-helix termed the ArfGAP1 lipid packaging sensor (ALPS) that specifically binds to positively curved membranes (Bigay et al., 2005). The ALPS is unstructured in solution but folds into an α-helix with a weakly charged polar face rich in serine and threonine residues and a hydrophobic surface that inserts into packing defects in positively curved membranes (Drin et al., 2007). The absence of basic residues on the polar face makes ALPS a membrane curvature sensor rather than a promoter. Amphipathic α-helices rich in basic residues promote membrane curvature since the strong electrostatic interactions counteract the energetic cost of spreading lipid molecules apart. Thus, ArfGAP1 displays curvature-dependent Arf1-GTP hydrolysis activity *in vitro* that ensures efficient Arf1 dissociation from the membrane and a subsequent vesicle uncoating (Bigay et al., 2003).

DAG also promotes ArfGAP1 activation and association with the Golgi apparatus possibly related to its positive membrane curvature-inducing properties (Antonny et al., 1997; Fernandez-Ulibarri et al., 2007; Asp et al., 2009). PtdOH phosphatases (PAPs) that produce DAG play a role in COPI vesicle transport (Asp et al., 2009). Two types of PAPs, cytosolic PAP1 and membrane-bound PAP2, are expressed in mammalian cells. Pharmacological experiments demonstrated that PAP1 is mainly responsible for DAG synthesis relevant to ArfGAP1 binding to Golgi membrane. ArfGAP2 and ArfGAP3 lack the ALPS motif, and as such, their activities and Golgi localization are strictly dependent on coatomer interaction (Kliouchnikov et al., 2009). Thus, ArfGAP2 and ArfGAP3 might have a constitutive role in COPI assembly, whereas ArfGAP1 is a lipid-sensitive terminator of Arf1 activity and uncoating. Upon GTP hydrolysis, Arf1 undergoes dimerization on the membrane that is needed for curvature generation and COPI vesicle biogenesis (Beck et al., 2008). Mutant Arf1 that lacks the ability to dimerize still recruits COPI and associates with membranes via the amphiphatic α-helix. However, dimerization deficient Arf1 did not support the generation of mature transport vesicles indicating that insertion of a pair of α-helices is an initial step in vesicle biogenesis.

BARS was originally identified as a substrate for ADP-ribosylation in brefeldin A-treated cells, and antagonized the tubulation-inducing effects of brefeldin A on the Golgi complex (Mironov et al., 1997). In addition to effects on membrane fission, a nuclear form of BARS is also a transcriptional co-repressor (Nardini *et al.*, 2003). BARS binds to PtdIns, PtdIns(4)P and PtdOH, but not to DAG (Yang *et al.*, 2008). It is probably recruited by binding to ArfGAP1, and participates in membrane tubulation at a late stage of vesicle fission that is dependent on high concentrations of PtdOH (Yang *et al.*, 2005; Yang *et al.*, 2008). The structural basis for membrane deformation by BARS has yet to be established but analysis of truncation mutants revealed that the C-terminal domain is the minimal domain required for inducing tubulation in conjunction with PtdOH (Yang *et al.*, 2008). Notably, depletion of PLD2 by siRNA induced accumulation of buds with constricted necks confirming a role for BARs in Golgi membrane fission (Yang *et al.*, 2008). As addition of PtdOH reversed this inhibition, PLD2 is required for final scission to generate COPI vesicles. More recently, BARS was demonstrated to activate PLD1 activity *in vitro* and during macropinocytosis suggesting it coordinates membrane fission events with localized changes in lipid composition (Haga, 2009).

2.3. Lipid Functions In Membrane Trafficking

Lipids that cause negative or positive curvature when they are asymmetrically distributed between the inner and the outer leaflets of a bilayer determine membrane shape. Such effects arise from the intrinsic shape of a lipid related to the relative volume occupied by the polar head group compared to the apolar fatty acyl moieties (Farsad and De Camilli, 2003).

Lipid molecules can be classified into three groups according to their space-filling properties (Corda *et al.*, 2002). PtdCho, PtdIns, phosphatidylserine (PtdSer), and phosphatidylglycerol are generally cylindrical in shape and are accommodated in planar bilayers. In contrast, DAG, PtdOH, unsaturated fatty acids, phosphatidylethanolamine (PtdEtn) and cholesterol are cone shaped (type II), conferring negative curvature to a membrane. Lysophospholipids have an inverted cone shape (type I) and induce positive curvature in membranes. These lipids are not homogeneously distributed across the secretory pathway, ultimately leading to intrinsic differences in membrane curvature and charge that are critical for membrane fission and fusion. Changes in lipid composition in the process of vesicle formation are mediated by events such as lipid translocation, metabolism, and transbilayer movement (flip-flop). In this section, we will focus on PtdOH, DAG, PtdIns, cholesterol, and SM, which play key roles in Golgi function and dynamics.

2.3.1. PtdOH, DAG, and Lyso-Phospholipids

2.3.1.1. PtdOH and DAG Synthesis and Interconversion

De novo biosynthesis of PtdOH occurs on the ER and the outer mitochondrial membrane. The first step is condensation of glycerol-3-phosphate and acyl-CoA to form lyso-PtdOH by glycerol-3-phosphate acyltransferase (GPAT) (Fig.2). Two GPATs in mammalian cells are localized in the ER (GPAT3) and the mitochondria (GPAT1) (Wendel *et al.*, 2009). Different forms of acyl-CoA synthases in the ER and mitochondria are thought to produce different pools of acyl-CoA that affect the molecular species of PtdOH (Wendel *et al.*, 2009). Expression of GPAT is stimulated by transcription factors such as sterol response element-binding protein (SREBP) and NF-Y (Ericsson *et al.*, 1997). The acylation of lyso-PtdOH by acyl-glycerol-3-phosphate acyltransferases 1 and 2 (LPAAT1 and 2) occurs in the ER. Thus much of the lyso-PtdOH produced in the mitochondria is transferred to the ER independent of carrier proteins (Das et al., 1992). PtdOH is also formed by DAG kinase and hydrolysis of phospholipids by phospholipase D (PLD). Although PtdOH is an important lipid component, it is short-lived and rapidly converted to DAG (Lev, 2006).

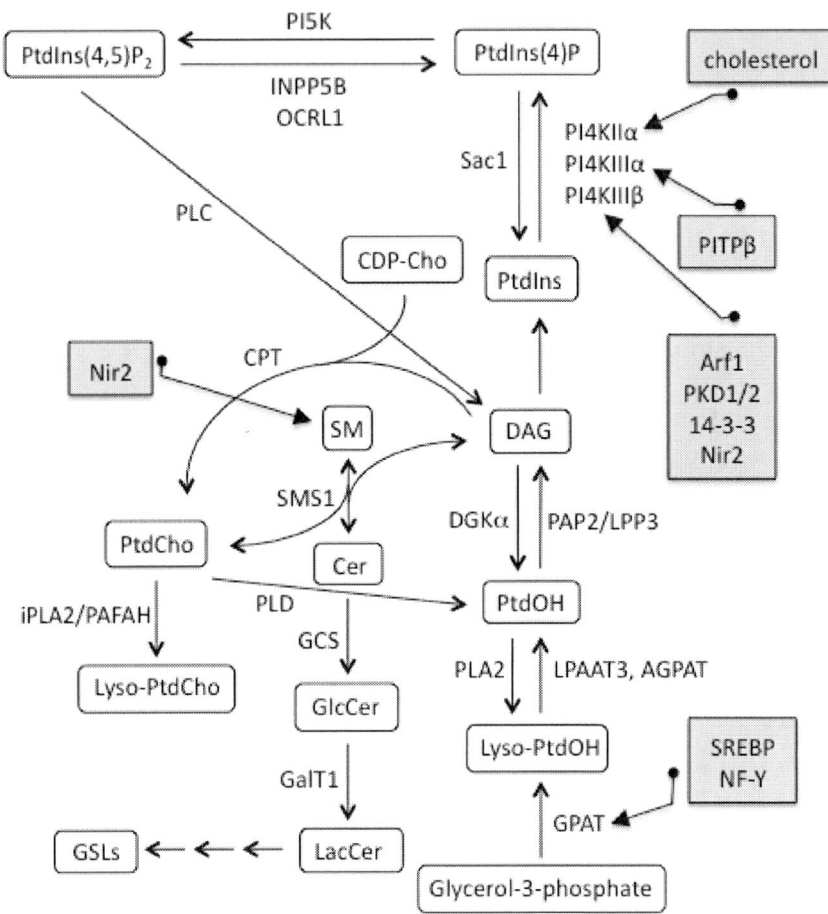

Figure 2. Metabolic pathways and interconversion of lipids involved in Golgi function.

DAG has a longer lifetime compared to PtdOH and exists at constant levels within the Golgi apparatus (Lev, 2006). DAG is produced during synthesis of SM from PtdCho and ceramide by SMS1, dephosphorylation of PtdOH by PtdOH phosphatase 2 (PAP2)/lipid phosphate phosphatase 3 (LPP3) and phospholipase C (PLC)-mediated hydrolyosis of PtdIns(4,5)P_2 (Kai et al., 1997) (Fig.2 and Fig.3). It is proposed that PLD contributes to DAG production in the PAP pathway by promoting the formation of PtdOH from PtdCho (Billah et al., 1989). Two subtypes (PLD1 and PLD2) localize within different parts of the Golgi where they have distinct functions in vesicular trafficking (Kim et al., 1999; Freyberg et al., 2002). As described in Section 2.2, PLD2 is a crucial component of COPI vesicle formation (Yang et al., 2008). The defining feature of the PLD enzymes is the HKD catalytic motifs

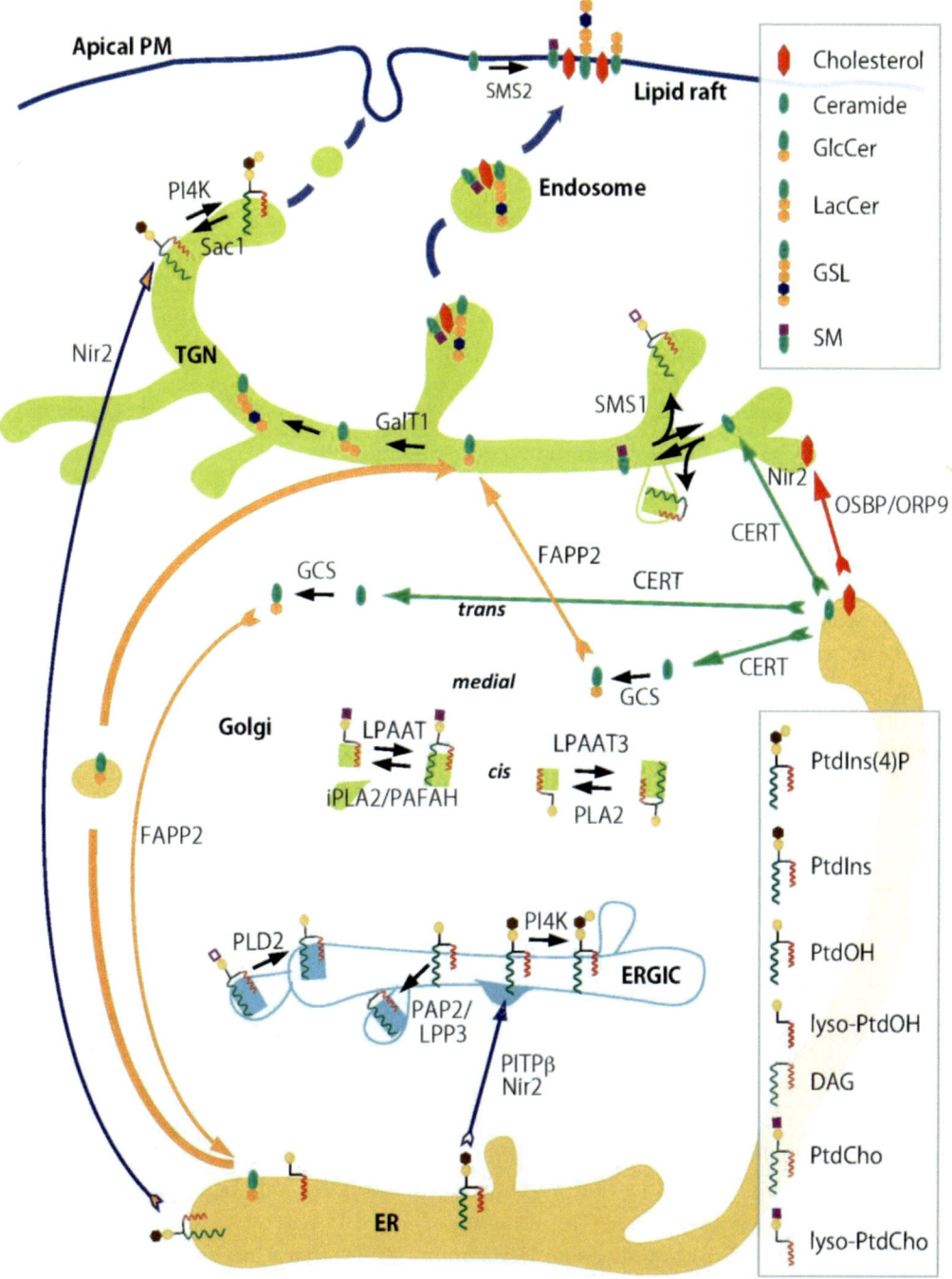

Figure 3. Biosynthesis and transfer of lipids involved in Golgi function. Symbols for each lipid are indicated in the box. PtdIns is produced in the ER and subsequently transferred by PITPs to sites of PtdInsP production. PtdIns(4)P serves to recruit protein effectors that are involved in vesicular and non-vesicular transport to and from the Golgi. DAG, PtdOH and lyso-PtdOH flip across and interconvert to condense at sites of membrane deformations. Cholesterol transferred from the ER is packaged together with SM and GSLs and coalesce into lipid raft for transport to the apical PM. Refer to the text for details.

(HXXXXKXD) (Sung et al., 1997). Other conserved regions of PLDs are the three lipid-binding phox (PX), PH and polybasic domains (Jenkins and Frohman, 2005). In addition, PLD1 has a unique conserved loop region that is a negative regulatory element (Sung et al., 1999). As a result, PLD1 exhibits low basal activity and is activated by many factors including PtdIns, Arf, Rho and PKC (Jenkins and Frohman, 2005), whereas PLD2 is constitutively active (Sung et al., 1999). PLD1 and PLD2 activities are both dependent on PtdIns(4,5)P$_2$ (Jenkins and Frohman, 2005; Riebeling et al., 2009).

The capability of Golgi membranes to produce DAG is countered by efficient clearance pathways. These include PtdCho synthesis by reversal of SMS1 (Section 2.3.3.3.) and the CDP-choline pathway (Fig.2 and Fig.3). The terminal steps in the CDP-choline pathways are catalyzed by an ER-localized choline/ethanolamine phosphotransferase (CEPT) and a Golgi-localized choline phosphotransferase (CPT), enzymes that utilize CDP-choline/ethanolamine and DAG (Vance and Vance, 2004). While the exact contribution of these enzymes to total PtdCho synthesis is unknown, CPT activity could attenuate DAG levels in the Golgi apparatus.

PYK2 N-terminal domain-interacting receptor (Nir2) controls local DAG levels at membrane fission sites. Nir2 is a member of a highly conserved PI-transfer protein family (PITP) (Lev et al., 1999) that translocates from the Golgi apparatus to lipid droplets under conditions of oleic acid treatment (Litvak et al., 2002; Lev, 2004). PITPs are defined by their ability to transport PtdIns and PtdCho between membranes (Cockcroft and Carvou, 2007). Amongst the many lipid-transfer proteins identified, only the PITPs demonstrate dual specificity for phospholipid binding. Class I and class III PITPs (CRAL/Trio family) that include Sec14-like proteins and other mammalian homologues are small soluble proteins (Allen-Baume et al., 2002; Goldstein et al., 2006b). Although mammalian Sec14-like proteins share homology with the yeast PITP sec14, their ability to transfer lipids has not been examined (Allen-Baume et al., 2002). On the other hand, class II PITPs (Nir family) are large proteins that have an N-terminal PI-transfer domain (amino acids 1-257) that is followed by six transmembrane domains, a 180-residue-long region with conserved DHDD residues and a long C-terminal region (Lev, 2004). Just after the PI-transfer domain is an acidic region containing a two phenylalanines in an acidic tract (FFAT) motif that binds vesicle-associated membrane protein-associated protein (VAP). The mechanism for Golgi localization of Nir2 is unknown. The DHDD residues constitute a metal-binding site often seen in phosphoesterase domains but its role in Nir2 function is unclear. The C-terminus region is involved in protein-protein interaction with the tyrosine kinase PYK2 (Lev et al., 1999). In a PI-bound state, Nir2 facilitates PtdIns(4)P production at the Golgi (Aikawa et al., 1999) (Section 2.2.2.), whereas its PtdCho-bound form negatively regulates the production of PtdCho via the CDP-choline pathway resulting in increased DAG (Litvak et al., 2005). Thus it is considered that Nir2 links PtdIns and DAG homeostasis at the Golgi via interaction with VAP at membrane contact sites to facilitate PI transfer (Peretti et al., 2008).

2.3.1.2. Roles for PtdOH and DAG in Vesicular Fission at the Golgi

Membrane curvature is influenced by the relative abundance of type I and II lipids such as PtdOH and DAG whose asymmetric distribution in the inner or outer leaflets of the lipid bilayer confer negative or positive membrane curvature. PtdOH is a unique fusogenic lipid that changes its shape in a calcium-dependent manner. It adopts a cylindical shape under conditions of low calcium concentration, but becomes conical shaped in the Golgi lumen where calcium concentration is approximately 0.3 mM (Pinton *et al.*, 1998; Kooijman *et al.*, 2003). Moreover, PtdOH can translocate across model membranes (Eastman *et al.*, 1991). Thus, PtdOH could translocate across Golgi membranes and induce negative membrane curvature and fusion in the presence of divalent cations. As outlined in Section 2.3.1.1., PtdOH is also rapidly converted to DAG at the Golgi by the action of PtdOH phosphatases. DAG induces negative membrane curvature upon accumulation in a lipid leaflet, which is speculated to facilitate membrane budding, fission and fusion (de Figueiredo *et al.*, 1999; Bard and Malhotra, 2006; Lev, 2006). Thus rapid interconversions between DAG, PtdOH and its precursor lyso-PtdOH, an inverted-cone shape, might facilitate formation of highly curved membrane intermediates at fission and fusion zones through coordinated changes in local membrane composition (Fig.3). In addition, DAG can spontaneously flip-flop across membranes due to its small, uncharged head group (Corda *et al.*, 2002). This affects the composition and curvature of both leaflets in the bilayer no matter where the conversion of PtdOH to DAG occurs. However, as the enzymatic activities that produce DAG are located on the cytoplasmic face of the membrane, accumulation of DAG would tend to promote negative curvature at the neck of budding vesicles (Bard and Malhotra, 2006) (Fig.3). Indeed, localized generation of DAG on the outer leaflet of the TGN may be a key event in generation of TGN-to-plasma-membrane carriers (TPC). Treatment with fumonisin B1 blocks TPC formation (Baron and Malhotra, 2002) by inhibiting ceramide synthase S1-6 (Wang *et al.*, 1991), thus depriving SMS1 of substrate and lowering DAG levels in the Golgi. Down-regulation of Nir2, a regulator of DAG homeostasis, also blocks the fission of TPC (Litvak *et al.*, 2005). However, it is unclear how the DAG regulatory networks are connected with TPC formation.

In addition, DAG acts as a second messenger that recruits and activates specific protein modulators that control membrane transport. Non-polarized HeLa cells use a protein kinase D (PKD)-dependent pathway to deliver basolateral proteins, but not apical proteins, to the PM via activation of a G protein-coupled receptor at the TGN, release of activated Gα and Gβγ subunits and DAG production (Yeaman *et al.*, 2004). DAG then recruits protein kinase Cη (PKCη) and PKD1 through interaction with one of its cysteine-rich zinc-finger domains (Baron and Malhotra, 2002; Diaz Anel and Malhotra, 2005). Subsequently, DAG-activated PKCη binds to the PKD1 PH domain and activates via phosphorylation within the activation loop (Diaz Anel and Malhotra, 2005). Several substrates with the PKD consensus phosphorylation motif (LXRXXS/T) have been identified. PKD1 phosphorylates PI4KIIIβ at the Golgi leading to increased production of PtdIns(4)P (Hausser *et al.*, 2005), an important lipid that maintains Golgi structure and secretory function (Section 2.2.). In addition to PKD1, DAG activates PKD2 and PKD3, which also localize to the Golgi apparatus (Yeaman *et al.*, 2004). However, functional redundancies with in this PKD family are not well understood.

2.3.1.3. Role of Lyso-PtdOH in Golgi Secretion

Several phospholipase A2 (PLA2) enzymes have been localized to the Golgi apparatus where they are implicated in formation of tubular membrane structures that mediate cargo transport (Fig.3). Initial studies using pharmacological inhibitors implicated PLA2 in tubular membrane formation in various organelles, including the Golgi apparatus (de Figueiredo *et al.*, 1998; de Figueiredo *et al.*, 1999). The group IV calcium-dependent phospholipase A2α (cPLA2α) translocates to the Golgi apparatus by a cargo- and calcium- dependent mechanism where it induces the formation of intra-cisternal tubules (San Pietro et al., 2009). Platelet activating factor acetylhydrolase (PAF-AH) 1β is a heterotrimerc phospholipase complex of catalytic PLA2 subunits α1 and/or α2, and a non-catalytic dynein regulator lissencephaly 1 (LIS1) β subunit. This complex tubulates Golgi membranes *in vitro*, and localizes to the Golgi apparatus and regulates its morphology. RNAi suppression or overexpression of catalytically active or inactive α subunits fragmented the Golgi apparatus by inhibition of tubule-mediated Golgi transport and secretion (Bechler et al., 2010). The calcium-independent iPLA$_{1\gamma}$ is also localized to the *cis*-Golgi and ER-Golgi intermediate compartment (ERGIC) where it is implicated in Golgi to ER retrograde transport pathway that did not involve COPI (Morikawa et al., 2009). The wedge-shaped lyso-lipids produced locally by these enzymes could increase positive membrane curvature and tubule formation probably in conjunction with other stabilizing proteins. This finding not only identify lyso-lipids as potential effectors of intra-Golgi transport but highlight the role of intra-cisternal tubules as conduits for cargo transfer.

The membrane modifying effects of lyso-PtdOH and other lyso-phospholipids is counteracted by the activity of a class of 9 acyl-CoA-dependent lyso-PtdOH acyltransferases (LPAATs) with activity toward a variety of lyso-phospholipids (Shindou and Shimizu, 2009) (Fig.3). The drug CI-976, an inhibitor of acyl-CoA-dependent cholesterol esterification, also inhibits a Golgi-associated LPAAT activity leading to increased Golgi membrane tubules and retrograde transport (Drecktrah et al., 2003). More recently, LPAAT3 was identified as the CI-976-sensitive lyso-PtdOH-specific acyltransferase that when silenced caused Golgi fragmentation in HeLa cells (Schmidt and Brown, 2009). LPAAT3 in found in both the Golgi and ER but is distinct from LPAAT 1 and 2 involved in de novo synthesis of PtdOH in the ER. Overexpression of LPAAT3 inhibits retrograde trafficking to the ER and tubulation of Golgi membranes *in vitro*. Mechanistically, LPAAT3 would consume negative curvature-inducing lyso-PtdOH and thus counter the effects of PLA2 that promote Golgi tubule formation (Fig.2). Whether other LPAATs have similar activity in the Golgi or other endomembranes has yet to be determined.

2.3.2. PtdIns-Polyphosphates

2.3.2.1. Roles of PtdInsP in Golgi Trafficking

PtdInsP have three important roles in membrane fusion events: 1) serving as landmarks for recruitment of proteins that influence tethering, docking or fusion, 2) direct regulation of the fusion machinery, and 3) altering properties, such as fluidity or curvature, of fusogenic membranes (Fig.3). Specific interactions between PtdInsP and proteins are dependent on modules such as pleckstrin homology (PH), GAT and epsin N-terminal homology (ENTH) domains that recruit soluble proteins to membrane surfaces. In addition to specific binding domains, non-specific interactions through electrostatic attraction are also involved.

PtdIns(4)P is found predominantly in TGN (De Matteis et al., 2005) where local synthesis and catabolism controls lipid-transfer and coat complex formation required for Golgi function. For example, PH domain proteins such as the four-phosphate adaptor protein (FAPP2) (Godi et al., 2004; D'Angelo et al., 2007; Halter et al., 2007; Cao et al., 2009), CERT (Hanada et al., 2003), oxysterol-binding protein (OSBP) and several OSBP-related proteins (ORPs), including ORP9 and ORP11 (Storey et al., 1998; Ngo and Ridgway, 2009), regulate cholesterol and sphingolipid metabolism and trafficking in the Golgi complex (Fig.3). Adaptor and coat complexes (AP-1, GGA, EpsinR) facilitate budding of clathrin-coated vesicles via direct binding with PtdIns(4)P (Hirst et al., 2003; Wang et al., 2003; Wang et al., 2007). Additionally, a type IV P-type ATPase required for phospholipid translocase (flippase) activity and transport vesicle budding from the TGN has shown to be a PtdIns(4)P effector in Saccharomyces cerevisiae (Natarajan et al., 2009).

In contrast, PtdIns(4,5)P_2 exists mainly in the PM but is distributed in approximately 25% of tubulovesicular structures and Golgi cisternae (Watt et al., 2002) where it is implicated in vesicle trafficking and maintenance of Golgi morphology. Treatment with 1-butanol to inhibit PtdOH production by PLD lead to a dramatic decrease in Golgi PtdIns(4,5)P_2 but not PtdIns(4)P, which correlated with fragmentation of the Golgi cisternae into a uniform population of vesicles (Sweeney et al., 2002). Washout of 1-butanol restored PtdIns(4,5)P_2 synthesis, Golgi structure and enabled limited vesicle release.

2.3.2.2. PtdInsP Metabolism in the Golgi Apparatus

PtdIns have a D-myo-Ins head group esterified to DAG through a phosphodiester bond at the C-1 position. PtdIns comprise 5-8% of total cellular lipids and are synthesized in the ER. PtdIns can be reversibly phosphorylated in a variety of combinations on the 3, 4, and 5 hydroxyl groups of inositol to yield PtdIns(3)P, PtdIns(4)P, PtdIns(5)P, PtdIns(3,4)P_2, PtdIns(3,5)P_2, PtdIns(4,5)P_2, and PtdIns(3,4,5)P_3. Typically, PtdInsP are present in low amounts in membranes compared to other phospholipids (Vanhaesebroeck et al., 2001; Lemmon, 2008). PtdIns(4)P and PtdIns(4,5)P_2 have been localized in the Golgi by microscopy techniques (Rusten and Stenmark, 2006). Details of PtdIns(4)P localization within the Golgi compartment and contribution of PtdIns(4) kinases have been identified using PtdIns(4)P-specific PH domains fused to a GFP reporter (Balla et al., 2005; Weixel et al., 2005).

PtdIns(4)P and PtdIns(4,5)P_2 content of the Golgi is regulated by Golgi-localized kinases and phosphatases, as well as PITPs that provide PtdIns to these kinases. Nir2 associates with PI4KIIIβ and enhances PtdIns(4)P production in the Golgi (Aikawa et al., 1999). In addition, a class I PITPβ resides in the Golgi (Cockcroft and Carvou, 2007). PITPβ consists of an eight-stranded β-sheet flanked by two long α-helices that form a hydrophobic cavity capable of shielding a single lipid molecule (Vordtriede et al., 2005). Gene silencing of PITPβ caused defective COPI-mediated retrograde transport from the Golgi to the ER (Carvou et al., 2010). Since anterograde traffic from the TGN and SM or GSL synthesis was not affected, PITPβ has been suggested to localize at the cis-Golgi and not affect the PtdIns(4)P pool at the TGN (Fig.3). In contrast, knockdown of Nir2, which regulates DAG levels at the TGN via the CDP-choline pathway, reduced anterograde transport due to defective Arf1 recruitment to the Golgi (Litvak et al., 2005) (Fig.3). As PITPβ does not affect the recruitment of Arf1, which is dependent on DAG for localization at the Golgi (Section 2.2.), it is suggested that PITPβ does not regulate DAG in the Golgi. Moreover, knockdown of PITPβ caused Golgi compaction and actin accumulation suggesting that PITPβ is involved in COPI-mediated retrograde transport through recruitment of COPI and modulation of cytoskeletal dynamics.

PI4KIIα and PI4KIIIβ synthesize PtdIns(4)P in distinct regions of the Golgi apparatus; PI4KIIα co-localized with the TGN and endosomes, whereas PI4KIIIβ was present in the cis/medial Golgi compartments (Weixel et al., 2005). Thus PtdIns(4)P is produced by unique kinases at the suborganellar level where it has distinct roles in Golgi function. PI4KIIα is involved in sorting via the AP-1/clathrin-dependent pathway (Fig.1). PI4KIIα RNAi decreases Golgi PtdIns(4)P levels and blocks AP-1/clathrin complex recruitment to the TGN (Wang et al., 2003). Although AP-1 recruitment to the Golgi is dependent on Arf1 (Godi et al., 1999), PI4KIIα localization to particular regions within the Golgi is independent of Arf1 (Wang et al., 2003). As the activity of PI4KIIα responds to cholesterol (Waugh et al., 2006) (Fig. 2), it is possible that PI4KIIα localizes preferentially to cholesterol-rich regions within the TGN. PI4KIIIβ is recruited to the Golgi apparatus in combination with the GTP-bound form of Arf1 and Ca^{2+}-binding protein neuronal calcium sensor-1 (NCS-1). It participates in AP-1/clathrin-dependent vesicle transport (Godi et al., 1999; Haynes et al., 2005; de Barry et al., 2006) (Fig.1). This is independent of other activities of Arf1, such as COPI recruitment and PLD activation (Godi et al., 1999). PI4KIIIβ is also activated by PKD1 and PKD2 mediated phosphorylation (Hausser et al., 2005), which induces 14-3-3 protein binding to the phosphorylation site thereby protecting PI4KIIIβ from phosphatase-mediated dephosphorylation (Hausser et al., 2006) (Fig. 2). This leads to a continuous supply of PtdIns(4)P required for transport vesicle biogensis at the TGN. In addition, PI4KIIIβ specifically binds the GTP-bound form of Rab11 independent of its kinase activity (de Graaf et al., 2004). PI4KIIIβ binding is essential for the localization of Rab11 in the Golgi complex, where it participates in TGN-to-PM trafficking (Chen et al., 1998). PtdIns(4)P synthesized by PI4KIIIβ is also required for the transfer of ceramide from the ER to the Golgi apparatus (Toth et al., 2006)(Section 2.3.3.4, Fig.3)

Sac1 is a PtdIns(4)P-phosphatase with catalytic activity towards PtdIns(3)P, PtdIns(4)P and PtdIns(3,5)P$_2$ that predominantly localizes in the Golgi and regulates PtdIns(4)P levels (Nemoto et al., 2000). It has a phosphatase domain with a consensus $CX_5R(T/S)$ motif that is also present in metal-independent protein and lipid phosphatases (Whisstock et al., 2002). Sac1 synchronizes the secretory pathway with cell growth. During serum starvation and in quiescent cells, Sac1 oligomerizes in the Golgi where it catabolizes PtdIns(4)P and slows constitutive cargo secretion (Blagoveshchenskaya et al., 2008). Conversely, in mitogen-stimulated cells, Sac1 localizes to the ER, resulting in increased Golgi PtdIns(4)P and sustained secretion. Sac1 shuttles between ER and Golgi in COPI and COPII vesicles via interactions through the C-terminal di-lysine motif (KxKxx) and the N-terminal leucine zipper motif, respectively. Additionally, p38 mitogen-activated protein kinase and extracellular signal-regulated kinase 1/2 (ERK) are required for Sac1 translocation (Blagoveshchenskaya et al., 2008).

PtdIns(4)P 5-kinase (PIP5K) activity is recruited to the Golgi in an Arf1-dependent manner (Godi et al., 1999; Jones et al., 2000a) (Fig.2), however, the molecular identity of the PIP5K isoform has not been determined.

OCRL (Lowe's oculocerebrorenal syndrome protein) homolog (OCRL1) is a Golgi-associated PtdIns(4,5)P$_2$ phosphatase that is detected in Golgi apparatus (Olivos-Glander et al., 1995; Dressman et al., 2000; Faucherre et al., 2003) and endosomes (Ungewickell et al., 2004; Choudhury et al., 2005; Erdmann et al., 2007) of various cultured cells. OCRL1 promotes adaptor/clathrin mediated endosome to Golgi transport by catalyzing conversion of PtdIns(4)P to PtdIns(4,5)P$_2$ at endosomes and the Golgi apparatus. OCRL1 directly interacts with clathrin heavy chain, and promotes clathrin assembly in vitro (Ungewickell et al., 2004; Choudhury et al., 2005; Erdmann et al., 2007). However, it is unknown whether ORCL1 acts in concert with other clathrin assembly proteins such as AP-1 and EpsinR. OCRL1 targeting to the Golgi is mediated through interaction with Rab1 and Rab6, while Rab5 and the Rab5 effector APPL1 promote OCRL1 targeting to the endosomes (Hyvola *et al.*, 2006; Erdmann *et al.*, 2007).

Inositol polyphosphate 5-phosphatase (INPP5B) is another inositol polyphosphate 5-phosphatase localizes to the Golgi that has 45% amino acid sequence identity to OCRL1 (Mitchell *et al.*, 1989; Jefferson and Majerus, 1995; Matzaris *et al.*, 1998). As INPP5B interacts with Rab5, functions of INPP5B might be partially redundant with OCRL1. However, since INPP5B does not interact with clathrin and is absent from clathrin-coated vesicles, it has been speculated to be involved in retrograde traffic from the Golgi to ER (Williams *et al.*, 2007).

2.3.3. Cholesterol and Sphingolipids

2.3.3.1. Roles of the Golgi Apparatus in Cholesterol and Sphingolipid Homeostasis

The mitochondria, nuclear envelope and ER contain <5% of their lipid mass as cholesterol. However, membranes of the secretory pathway become increasing enriched in cholesterol, culminating with the PM, which contains up to 90% of total cellular cholesterol (Lange *et al.*, 1989; Lange, 1991; Warnock *et al.*, 1993). Cholesterol in the Golgi apparatus follows an enrichment gradient from the *cis* and *medial* cisternae to the TGN (Orci *et al.*, 1981). Disrupting this gradient by depletion or repletion of cholesterol leads to defective Golgi function and structure. For example, depletion of cellular cholesterol interferes with the generation of negative membrane curvature in the luminal leaflet of secretory vesicles budding from the TGN and reversible blockage of secretory vesicle biogenesis (Wang *et al.*, 2000a). Depletion of ER cholesterol reduces lateral mobility of membrane proteins and prevents ER-to-Golgi transport (Ridsdale *et al.*, 2006; Runz *et al.*, 2006). On the other hand, increasing cellular cholesterol content is shown to trigger fragmentation of the Golgi apparatus that is reversed when excess cholesterol is removed. (Ying *et al.*, 2003). Actin is involved in the process of dispersal and reformation of the Golgi in response to overload and removal of excess cholesterol (Ying *et al.*, 2003). Addition of jasplakinolide, an actin-stabilizing chemical, inhibits dispersal of cholesterol-induced fragmented Golgi apparatus. Conversely, addition of cytochalasin D, an actin-disrupting agent, inhibited reformation of the Golgi apparatus after removal of excess cholesterol. Golgi cholesterol content and distribution is maintained by complex and poorly understood interaction between vesicular and protein-mediated transport pathways (Section 2.3.3.2.).

Cholesterol physically associates with sphingolipids in detergent-resistant domains (DRM) or 'lipid rafts', the assembly of which occurs in the Golgi apparatus (Heino *et al.*, 2000; Ridgway, 2000; Coskun and Simons, 2010) (Fig.3). During a recent Keystone Symposium the following definition for lipid rafts was adopted: "Membrane rafts are small (10-200 nm), heterogeneous, highly dynamic, sterol and sphingolipid-enriched domains that compartmentalize cellular processes. Small rafts can sometimes be stabilized to form larger platforms through protein-protein and protein-lipid interactions." One of the most important characteristics of lipid rafts is the inclusion or exclusion of proteins (Simons and Toomre, 2000). Those associated with lipid rafts include glycosylphosphatidylinositol (GPI)-anchored proteins, doubly acylated Src-family kinases and α-subunits of heterotrimeric G proteins, cholesterylated and palmitoylated proteins (Hedghog) and select transmembrane proteins (Brown and London, 1998; Hooper, 1999; Resh, 1999; Rietveld *et al.*, 1999). Raft lipid composition impacts on associated cell signaling pathways and other membrane associated activities (Simons and Toomre, 2000), in some cases as a result of changes in Golgi lipid metabolism. For instance, the cholesterol efflux pump ABCA1 that mediates initial steps in the reverse cholesterol transport pathway is localized in Lubrol-resistant DRMs and non-raft membranes (Klappe et al., 2009). Activity of ABCA1 is affected by sphingolipid metabolic pathways situated in the Golgi apparatus. In LY-A cells that have a missense mutation in CERT, the SM content is 65% compared to LY-A cells expressing wild–type CERT (Nagao et al., 2007). Decreased cellular SM content increased ABCA1-mediated cholesterol efflux 1.65-fold, an effect that was reversed by exogenous addition of SM. On the other hand, cholesterol decreased ATPase activity of ABCA1 when added to a reconstituted liposome (Takahashi et al., 2006). ABCA1 also modulates the membrane environment by promoting redistribution of SM and cholesterol, which results in an expanded non-raft membrane fraction. Upon expression of ABCA1 in BHK cells, SM, cholesterol at the PM became less resistant to cold Triton X-100 extraction (Klappe et al., 2009).

Vesicles from the TGN are characterized by a high content of GSLs, which have an inverted-cone shape that promotes positive membrane curvature. This lipid shape consideration is largely negated by the preferential physical interaction with cholesterol, which has the opposite configuration. GSLs and SM are packaged together with cholesterol in TGN-derived secretory vesicles and transported to the PM as an early event in lipid raft formation (Heino et al., 2000) (Fig.3). In polarized cells, apically sorted proteins become associated in the TGN with nascent lipid rafts in a cholesterol-dependent manner (Roper et al., 2000). Two apical membrane proteins, prominin and placental alkaline phosphatase, which differ in solubility in two different non-ionic detergents, reside in distinct PM membrane microdomains. Depletion of cholesterol differentially affected translocation of the two proteins from the TGN to lipid rafts at the apical surface of cells.

2.3.3.2. Cholesterol Flux and Biosynthesis

Eukaryotic cells derive their cholesterol by *de novo* synthesis or uptake from circulating lipoproteins (Brown and Goldstein, 1997). Cholesterol is synthesized in the ER by the mevalonate pathway, which is under negative feedback control by the rate-limiting enzyme HMG-CoA reductase (HMG-CoAR) and sterol regulatory element binding proteins (SREBP) 1 and 2. SREBPs are transcription factors that reside in the ER but are transported to the Golgi complex under sterol-limiting conditions by COPII vesicles. In the Golgi apparatus SREBPs undergo two proteolytic processing steps to release the N-terminal fragment, a basic helix-loop-helix leucine zipper transcription factor that is imported into the nucleus, where it activates transcription of genes involved in synthesis of cholesterol and fatty acids (Goldstein et al., 2006a). Conversely, accumulation of cholesterol in the in the ER above a 5% threshold inhibits the ER-Golgi transport of SREBP (Radhakrishnan et al., 2008). Cholesterol, but not oxysterols, induces a conformational change in the chaperone SREBP cleavage-activating protein (SCAP) causing it to bind to another ER membrane protein, insulin-induced gene (INSIG), resulting in SREBP retention within the ER (Brown et al., 2002; Adams et al., 2004). Oxysterols inhibit SREBP processing by binding to INSIG and enhancing interaction with the SREBP/SCAP complex (Fernandez-Ulibarri et al., 2007; Radhakrishnan et al., 2007). Additionally, INSIG promotes proteasomal degradation of HMG-CoAR by binding to its sterol-sensing domain when cholesterol levels are high (Sever et al., 2003; Song and DeBose-Boyd, 2004). SREBPs are also involved directly and indirectly in regulation of phospholipid and fatty acid metabolism (Bennett *et al.*, 1995; Lagace *et al.*, 2000; Ridgway and Lagace, 2003). SREBPs do not regulate the expression of genes involved in SM synthesis, however SM has a key role in the regulation of SREBP by sequestering cholesterol. CHO cells defective in serine palmitoyltransferase activity and SM synthesis are acutely sensitive to sterol suppression of SREBP (Worgall *et al.*, 2004). Similarly, depletion of SM by pharmacological inhibitors or exogenous sphingomyelinase suppresses SREBP processing, transcriptional activity and cholesterol synthesis (Porn and Slotte, 1990; Scheek *et al.*, 1997; Worgall *et al.*, 2004). Increasing SM synthesis by providing precursors of SM synthesis or reducing SM levels by hydrolysis with exogenous sphingomyelinase, increases and decreases SREBP and gene transcription, respectively. Mechanistically this involves the sequestration or release of cholesterol from PM raft domains, leading to increased or decreased cholesterol influx to the ER.

Exogenous cholesterol is obtained by receptor-mediated uptake of cholesteryl ester-rich chylomicron remnants and LDL. Following uptake of these lipoproteins by LDL and chylomicron remnants receptors, cholesteryl esters are hydrolyzed by acid lipase to provide unesterified cholesterol for membrane, steroid hormone, and bile acid synthesis. Unesterified cholesterol exits the endosomes via a Niemann-Pick C1 and C2 (NPC1 and NPC2)-dependent pathway before maturation to lysosomes (Sturley *et al.*, 2004). Mutations in NPC1 account for 95% of the cases of Niemann-Pick disease, which is characterized by aberrant cholesterol accumulation in the late endosomes. The remaining 5% of cases are caused by mutations in the NPC2 gene (Mukherjee and Maxfield, 2004; Scott and Ioannou, 2004). NPC1 is a protein of late LE/L that has thirteen predicted transmembrane domains, five of which constitute a sterol-sensing domain similar to SCAP (Altmann *et al.*, 2004). NPC1 does not bind INSIG (Ko *et al.*, 2001; Ohgami *et al.*, 2004) but binds 25-OH and cholesterol via different cytoplasmic and lumen domains. On the other hand, NPC2 is a 132 amino acid soluble, glycosylated protein that binds cholesterol but not oxysterols with μM affinity (Cheruku *et al.*, 2006; Xu *et al.*, 2008). NPC2 transfers cholesterol between phospholipid vesicles in a lysobisphosphatidic acid (lysobis-PtdOH)-dependent manner (Cheruku *et al.*, 2006; Xu *et al.*, 2008). NPC1 and NPC2 act together to promote cholesterol efflux from the endosome/lysosome (LE/L) to the PM and ER. A model has been proposed wherein NPC2 binds and transfers cholesterol to NPC1 at the limiting membrane of the LE, the sterol moves across the bilayer and is released to other acceptors (Wang *et al.*, 2010). Cholesterol released from the LE appears in the PM and ER, where it is re-esterified by ACAT1 and 2 (Joyce *et al.*, 2000). Recently, cell fractionation studies showed that LDL-derived cholesterol first appears in the TGN prior to transport to other sites (Urano *et al.*, 2008). Transport to the TGN was reduced in NPC cells lacking NPC1 and was dependent on the TGN-specific SNAREs VAMP4, syntaxin 6 and syntaxin 16. Thus, the Golgi apparatus is a primary redistribution site for lipoprotein-derived cholesterol.

2.3.3.3. Sphingolipid Biosynthesis

SM and GSLs are essential components of mammalian cell membranes and are enriched on the outer leaflet of the PM. Sphingolipids have a common a hydrophobic ceramide moiety composed of N-acylated sphingosine or, in the case of dehydroceramide, sphinganine (Breathnach, 2001). Ceramide is synthesized in the ER and subsequently transferred to the Golgi where it is converted to complex sphingolipids (Yamaji *et al.*, 2008) (Fig. 3). Glucose or galactose is linked to the primary hydroxyl group of ceramide through a β-glycosidic bond, thereby giving rise to glucosylceramide (GlcCer) and galactosylceramide (GalCer), respectively. Linkage of phosphorylcholine produces SM. Further additions of monosaccharides and sulfate groups to GlcCer give rise to a broad range of complex GSLs (http://www.sphingomap.com). This biosynthetic process is facilitated by the transfer of lipid intermediates from the ER to the Golgi and between Golgi compartments (Fig.3). GlcCer synthase localizes to cytoplasmic face of the *cis-medial* Golgi and pre-Golgi compartments (Halter *et al.*, 2007). The active site of GalCer synthase resides in the lumen of ER (Sprong *et al.*, 1998). The glycosyltransferases that convert GlcCer to complex GLSs reside in late Golgi compartments and have active sites situated on the luminal surface (Wennekes *et al.*, 2009), necessitating the transport of intermediates between and across Golgi membranes. Two models have been suggested for this transport pathway (Fig.3). In the first, GlcCer is synthesized on the cytosolic side of the early Golgi by UDP-Glc:glucosylceramide synthase (GCS) (Coste *et al.*, 1986; Futerman and Pagano, 1991; Jeckel *et al.*, 1992), and subsequently transferred to the *trans*-Golgi and flopped across the bilayer into the lumen where it is converted to lactosylceramide (LacCer) and other complex GSLs. The *cis*- to *trans*-Golgi transport of GlcCer is dependent on the GlcCer transport protein FAPP2 (D'Angelo *et al.*, 2007) (Section 3.1.2). Silencing FAPP2 by RNAi reduces the conversion of GlcCer to LacCer and to complex GSLs. GlcCer crosses Golgi membranes by the ATP-independent, bidirectional transporter P-glycoprotein (PgP, multidrug resistant transporter ABCB1) (Buton *et al.*, 2002). PgP is an ATP-dependent flippase for a variety of simple GSLs *in vitro*, including GlcCer, in a reconstituted proteoliposome assay (Eckford and Sharom, 2005). However, PgP cannot flip GSL derivatives larger than two sugar residues, suggesting alternative mechanisms. In the second model, GlcCer is proposed to be synthesized primarily in the *trans*-Golgi, transported back to the ER by FAPP2, flopped across the ER membrane and delivered by vesicular mechanism to the late Golgi for conversion to GSLs (Halter *et al.*, 2007). Treatment with brefeldin A, which induces ER fusion with the Golgi, increases synthesis of LacCer indicating that GlcCer flops more efficiently across the ER membrane compared to the Golgi. This non-linear, retrograde pathway relies on CERT for ceramide delivery to GCS in the late Golgi.

Having arrived at the luminal leaflet of the *trans*-Golgi, the biosynthesis of GSLs continues with the synthesis of LacCer by GalT1 (Wennekes *et al.*, 2009). LacCer is extended sequentially at either the 3- or 4-O-position in a stepwise fashion. The nature of the glycosylation reactions depend on the expression patterns of glycosyltransferases in a given cell type (Hakomori, 2000). Switching of GSL series occurs during differentiation. In erythroleukemia K562 and HEL cells, ganglio-series GSLs are predominant, but are decreased and replaced by lacto-series GSLs during differentiation. Synthesis of globo-series GSL is induced upon further differentiation. Tumors have specific GSL antigens that are not expressed in normal cells. Certain anti-GSL IgG mAbs were found to be useful for tumor growth suppression. Catabolism of complex GSLs is a stepwise process that takes place in endosomes and lysosomes (Wennekes *et al.*, 2009).

A family of choline and ethanolamine phosphotransferases (SMS1, SMS2 and SMSr) that synthesize SM and ceramide phosphoethanolamine (CPE) were identified by a database search for proteins containing a C3 motif shared by previously characterized LPPs. Mammalian SMSs are integral membrane enzymes with six membrane-spanning domains and four conserved motifs (D1-D4), of which D3 and D4 correspond to the C2 and C3 motifs in LPPs. Similar to LPPs, D3 and D4 are oriented towards the same side of the membrane and could be a part of the catalytic site responsible for liberating phosphocholine from PtdCho during SM synthesis. SMS1 in the *trans*-Golgi/TGN and SMS2 is PM both catalyze the reversible conversion of PtdCho and ceramide into SM and DAG (Huitema *et al.*, 2004) (Fig.2, Fig.3). Overexpression and RNAi studies indicate that SMS1 is an important determinant of cellular DAG and ceramide. SMS1 activity is regulated primarily by ceramide delivery from the ER by the ceramide transport protein CERT (Collins and Warren, 1992; Moreau *et al.*, 1993; Hanada *et al.*, 2003) (Section 3.1.2.). In turn, CERT transport activity is regulated by phosphorylation by PKD at serine 132, which triggers phosphorylation at adjacent serine and threonine residues by casein kinase I and inhibits ceramide transport activity by retention in the ER (Kumagai *et al.*, 2007). Since DAG activates PKD, this creates a negative feedback loop where phosphorylation of CERT decreases SMS1 activity, DAG generation and PKD activity. Indeed, overexpression of CERT or a phosphorylation-deficient mutant increased PKD activity and secretion in HEK293 cells.

SMS2 is a bi-functional enzyme with SM and CPE synthase activity (Ternes *et al.*, 2009), while SMS1 and SMSr are mono-functional enzymes with SM and CPE activity respectively. SMS2 is located at the PM where it resynthesizes SM to terminate ceramide signals produced following SM hydrolysis. SMS2 is involved in regulation of apoptosis by controlling the SM levels at the PM specifically in raft domains, which affects the cell surface expression of TNF-R and TLR-4 (Ding *et al.*, 2008). However, SMS2 is also detected in the Golgi apparatus and implicated in *de novo* synthesis of SM (Huitema *et al.*, 2004; Li *et al.*, 2007; Tafesse *et al.*, 2007). SMS2 also regulates DAG formation at the Golgi and thereby affects DAG-dependent localization of PKD to the Golgi (Villani *et al.*, 2008).

The related SMSr catalyzes the synthesis of CPE from ceramide and PtdEtn in the ER. CPE is an abundant sphingolipid in insects and nematodes where it is involved in regulation of sterol metabolism (Jacob and Kaplan, 2003; Rao *et al.*, 2007). CPS is also found in mammalian cells but at low abundance. Inhibition of SMSr activity caused disruption of ER exit sites and the Golgi apparatus in HeLa cells, not by depletion of ceramide phosphoethanolamine but by increasing ER ceramide. This suggests that SMSr negatively regulates ceramide levels in the early secretory pathway by inhibiting ceramide synthesis or increasing degradation.

2.3.4. Lipid Transport Proteins Involved in Cholesterol and Sphingolipid Metabolism

Although the vesicular and tubular carriers that move cargo in the Golgi apparatus are composed of phospholipids, sterols and sphingolipids, it is apparent that bulk membrane transport is not sufficient to support localized and specialized synthesis of Golgi lipids, notably sphingolipids and cholesterol. This is achieved by the lipid carrier proteins FAPP, CERT and members of the OSBP family that localize to the Golgi apparatus and ER via protein-lipid interactions. These dual membrane interactions coupled with high affinity lipid binding pockets are suggestive of site-specific, vectoral transport proteins.

The PH domains of FAPP, CERT and OSBP target these proteins to the Golgi apparatus via PtdIns(4)P binding (Lagace *et al.*, 1997; Godi *et al.*, 2004; Kumagai *et al.*, 2007) (Fig.3). Recent, NMR-based solution structures of the free, micelle- and PtdIns(4)P-bound FAPP1-PH domain showed that it contacts PtdIns(4)P through residues in the β1- β2 loop (Lenoir *et al.*, 2010). The insertion of the FAPP2 PH domain formed a wedge that penetrated the bilayer and induced positive curvature and membrane tubulation. This wedge conformation is highly conserved across the FAPP family, CERT, and OSBP family proteins, suggesting they all tubulate Golgi membranes by a similar general mechanism. Coincidental binding of these PtdIns(4)P-specific PH domains to Arf1 ensures highly specific organelle localization (Lenoir et al., 2010).

CERT and some OSBP subfamilies have FFAT motifs (EFFDAxE) that are responsible for ER localization through direct binding to vesicle-associated membrane protein (VAMP)-associated protein (VAP) (Wyles et al., 2002; Kawano et al., 2006). VAPs are type II membrane proteins that are conserved from yeast to human (Loewen et al., 2003; Lemmon, 2008). Three VAP isoforms have been identified in humans: VAP-A, VAP-B, and VAP-C (a splicing variant of VAP-B). Although VAP localization to other organelles is reported (Lapierre et al., 1999; Skehel et al., 2000), they are generally localized in the ER. VAP consist of three domains: an N-terminal immunoglobulin-like β-sandwich fold consisting of seven β-strands and one α-helix that is similar to the nematode major sperm protein (MSP); a variable central coiled-coil domain that resembles VAMPs and other SNARE proteins; and a C-terminal transmembrane domain containing a putative dimerization motif (GxxxG) (Lev et al., 2008). VAPs are involved in vesicular lipid transport, lipid metabolism, regulation of ER structure, and the unfolded protein response by interacting with FFAT motifs through the MSP-like domain (Lev et al., 2008). Crystallographic and NMR analysis have revealed that five of the six conserved residues of the FFAT motif are required for stable complex formation with the MSP-like domain by a combination of electrostatic, hydrophobic, and hydrogen-bond interactions (Kaiser et al., 2005; Furuita et al., 2010). The region of OSBP proximal to the FFAT domain has an overall negative charge that forms electrostatic interactions with a positively charged surface containing the MSP-like domain.

2.3.4.1. Glycolipid Transfer by FAPP

The concept of glycolipid synthesis in the Golgi was one of linear path for substrate delivery by vesicular transport from the early to the late Golgi cisternae where the various glycosyltransferases would sequentially add sugar moities to ceramide and GlcCer. The identification of the GlcCer and ceramide transport proteins FAPP2 and CERT has lead to revision of this model to include site-to-site lipid transport. FAPP1 and FAPP2 were initially identified as effectors of Arf1 and PtdIns(4)P that regulate cargo transfer from the Golgi to PM (Godi et al., 2004) (Fig.3). They are recruited to the Golgi through their PtdIns(4)P-specific PH domains. FAPP2 contains a glycolipid-transfer protein (GLTP) homology domain that has GlcCer transfer activity. Two different models have been proposed for FAPP2 delivery of GlcCer in the ER and Golgi pathway; transfer from the *cis*- to *trans*-Golgi (D'Angelo et al., 2007) and from the *trans*-Golgi to the ER or PM (Halter et al., 2007). The integration of FAPP2 activity with GSL synthesis in the late Golgi compartments is described in Section 2.3.3. As FAPP1 lacks GLTP domain, it is considered not to have GlcCer transfer activity.

In addition to GlcCer transport, FAPP2 has a role in apical transport through its membrane tubulating activity (Cao et al., 2009). When expressed in Cos-7 or MDCK cells, FAPP2 was present in tubules forming from the *trans*-Golgi (Godi et al., 2004; Cao et al., 2009). Moreover, FAPP2 induced growth of membrane tubules from lipid sheets composed of PtdCho, PtdIns(4)P and GlcCer in a PH domain dependent-manner (Cao et al., 2009). Dimerization of FAPP2 has been demonstrated, and a low-resolution structural analysis of the full-length FAPP2 protein shows that the GLTP and PH domain are located at opposite ends of the FAPP2 dimer (Cao et al., 2009). Together with results showing that FAPP1-PH domain also can penetrate into lipid monolayers and decrease the membrane pressure (Stahelin et al., 2007), it is apparent that membrane binding and tubulation activities reside in the conserved FAPP1 and FAPP2 PH domain.

2.3.4.2. Ceramide Transfer Protein

CERT was originally identified as GPBPD26, a splice variant of Goodpasture antigen-binding protein (GPB), a kinase that binds and phospholylates the non-collagenous C-terminal (NC1) region of the α3 chain of collagen IV, the antigen in autoimmune Goodpasture disease (Raya et al., 2000). The exact function of GPBP remains elusive, whereas the GPBPD26 variant CERT is involved in SM synthesis by mediating non-vesicular transport of ceramide from the ER to the Golgi apparatus (Hanada et al., 2003). CERT, a member of the steroidogenic acute regulatory (StAR) protein-related lipid-transfer (START)-domain family (Soccio and Breslow, 2003), contains a C-terminal ceramide binding START motif, a FFAT motif and a N-terminal PH domain. START domains are protein modules of ~230 amino acids similar to the founding member of this family, StAR (Ponting and Aravind, 1999). The human genome encodes fifteen START domain proteins that bind a variety of lipid ligands such as cholesterol, PtdCho and ceramide. A recent study solved the crystal structure of the CERT START domain in complex with ceramide analogs with varying acyl chain length (Kudo et al., 2010). Only one ceramide molecule can be accommodated in the CERT START domain, where it forms a hydrogen bond network with specific amino acid residues at the far end of the cavity. As there is no extra space to accommodate an additional bulky group at the C1 position of ceramide, CERT does not bind SM.

Localization and activity of CERT is controlled by a phosphorylation-dephosphorylation cycle. Dephosphorylated CERT is in an 'open' conformation that facilitates PH domain interaction with PtdIns(4)P in the *trans*-Golgi/TGN. There it releases ceramide for SM synthesis catalyzed by SMS1 (Kumagai et al., 2007) (Fig.3). DAG, another product of SMS1, recruits PKD that phosphorylates CERT on a serine that then primes the subsequent phosphorylation by casein kinase I γ2 (CKIγ2) of 9 sites in the serine-rich motif (Hanada et al., 2009). Hyperphosphorylated CERT has reduced affinity for PtdIns(4)P and dissociates from the Golgi. Dephosphorylated CERT interacts with VAP in the ER and acquires ceramide for transport to the Golgi apparatus. Reloading of CERT with ceramide also involves CERT dephosphorylation by phosphatase 2Cε (PP2Cε), an integral membrane protein located in the ER (Saito et al., 2008). CERT transport of ceramide from the ER to Golgi could involve a sequence of discrete dissociation/binding steps as described above. However, CERT has the potential to simultaneously contact both donor and acceptor membranes via PH and FFAT domains and thus transport could proceed at membrane contact sites between these organelles (Levine, 2004; Hanada et al., 2009).

2.3.4.3. Golgi Cholesterol Transfer by OSBPs

OSBP was originally identified as a cytosolic receptor for oxysterols, oxygenated derivatives of cholesterol (Kandutsch and Shown, 1981). Eleven other OSBP-related proteins (ORPs) have since been identified in the human and other mammalian genomes with numerous variants generated by alternate promoter and splice site usage (Ngo et al., 2010). OSBP and ORPs have an OSBP homology (OH) domain that is highly conserved across eukaryotic genomes (Levanon *et al.*, 1990; Jiang *et al.*, 1994; Alphey *et al.*, 1998). OH domains are ~400 amino acid residues with a conserved signature motif (EQSHHPP), and bind cholesterol and/or oxysterols with affinities in the low to mid nM range (Ridgway *et al.*, 1992; Suchanek *et al.*, 2007; Wang *et al.*, 2008). The structure of Osh4, a yeast homolog of ORP, has been solved in complex with several sterols and oxysterols (Im *et al.*, 2005). A single sterol molecule binds in the hydrophobic tunnel of 19-strand β-barrel with the 3-OH group orientated toward the bottom. The entrance of the tunnel is blocked by a flexible N-terminal lid and surrounded by conserved basic residues. Osh4 undergoes a conformational change involving the opening and closing of this flexible lid over the mouth of binding cavity (Canagarajah *et al.*, 2008). Side-chain oxysterols (such as 25-hydroxycholesterol) stabilize the closed conformation of the lid by direct hydrogen bonding with lid residues. In addition to the OH domain, most ORPs contain PH and FFAT domains that interact with PtdInsPs and VAP, respectively. However, some ORP subfamilies lack FFAT and PH domains or have C-terminal transmembrane domains that mediate ER localization. Growing evidence indicates OSBP involvement in diverse cellular functions such as sterol signaling, sterol transport, regulation of lipid metabolism and cytoskeletal organization (Wang *et al.*, 2005; Bowden and Ridgway, 2008; Zerbinatti *et al.*, 2008).

A subset of OPRs, including OSBP, ORP9, and ORP11, interact with the Golgi and ER, and are involved in binding and/or transfer of cholesterol (Ridgway *et al.*, 1992; Ngo and Ridgway, 2009; Zhou *et al.*, 2010) (Fig.3). OSBP mediates the activation of SM synthesis in response to change in cellular cholesterol and/or oxysterol levels (Storey *et al.*, 1998). Overexpression of OSBP enhances synthesis of SM in response to 25-hydroxycholesterol (25-OH) (Lagace *et al.*, 1999). The mechanism involves oxysterol or cholesterol binding to OSBP and increased CERT translocation from the ER to Golgi apparatus (Perry and Ridgway, 2006). As OSBP and CERT do not interact directly, additional factors are involved in OSBP-mediated CERT translocation. Recently we discovered that OSBP activates the sterol-sensitive PI4KIIα in the TGN, leading to increased PtdIns(4)P, CERT recruitment and increased SM synthesis (Banerji, *et al.* 2010). Thus, OSBP is a key regulator that links cholesterol and SM metabolism, which in turns could affect lipid raft assembly in the late Golgi (Section 2.3.1.).

ORP9 also localizes to the Golgi apparatus but is not involved in SM synthesis nor does it translocate from the ER to Golgi in response to cholesterol and/or oxysterol (Wyles *et al.*, 2002; Ngo and Ridgway, 2009). Instead, depletion of ORP9 by RNAi knockdown causes fragmentation of the Golgi and increased cholesterol in an endosomal compartment that stained with filipin. This was correlated with inhibition of ER-Golgi vesicular transport (Ngo and Ridgway, 2009). Recently ORP9 was shown to physically interact with ORP11 so it is possible that these two ORPs are involved in a common pathway affecting Golgi or post-Golgi cholesterol homeostasis and organelle structure (Zhou *et al.*, 2010).

Phosphorylation sites were identified in OSBP and ORP9 that regulate function. Acute and long-term depletion of PM cholesterol promotes dephosphorylation at 3 serine residues adjacent to the sterol-binding domain and increased Golgi localization of OSBP (Storey *et al.*, 1998; Mohammadi *et al.*, 2001). This serine-rich site is similar to the site in CERT that regulates localization and is sensitive to PM cholesterol levels, however in the case of OSBP the initiating kinase is unknown. Phosphorylation of OSBP on serine 242 by PKD results in Golgi fragmentation and decreased transport of VSV-G protein to the PM (Nhek *et al.*, 2010). The presence of these two sites in CERT and OSBP that are controlled by cholesterol levels and PKD suggests a coordinated mechanism to regulate SM and cholesterol delivery and synthesis at the Golgi for raft assembly.

Whether phosphorylation affects ORP9 localization to the Golgi remains unclear. Although ORP9 is a substrate of PDK-2, thereby negatively regulating Akt phosphorylation in a competitive manner, mutation of the PDK-2 phosphorylation site had no effect on Golgi localization (Lessmann *et al.*, 2007).

3. PROTEINS INVOLVED IN MEMBRANE TETHERING AND FUSION

The fusion of vesicles or tubulovesicular structures with acceptor membranes involves a complicated interplay between lipid and protein factors. The stalk model for membrane fusion posits a series of intermediate structures catalyzed by fusion proteins (Chernomordik and Kozlov, 2008). First, the outer leaflets of the two bilayers merge to form a hemifusion intermediate. Then a fusion pore opens and the inner leaflets merge. Studies of protein-free lipid bilayers have revealed conditions required for membrane fusion. Positive curvature lipids, such as lysophosphatidylcholine (lyso-PtdCho), and negative curvature lipids, such as PtdEtn and fatty acids, inhibit and promote hemifusion, respectively, at the contact site between the two lipid bilayers (Chernomordik *et al.*, 1995). This indicates that hemifusion involves formation of lipid intermediates of negative curvature. Conversely, lyso-PtdCho and PtdEtn promote and inhibit pore opening (Chernomordik *et al.*, 1995). Experiments demonstrating that small diameter liposomes are the most fusogenic indicates that membrane tension is critical for fusion (Malinin *et al.*, 2002). Although the specific mechanisms by which proteins promote hemifusion and pore opening remain elusive, several proteins that are involved in these processes have been studied.

Tethering factors and SNAREs control Golgi vesicle fusion and, tethering between two Golgi cisternae for linking of the Golgi ribbons (Fig.1). Membrane tethering and fusion of transport vesicles mediated by tethering factors and SNAREs can be described in three steps. (1) A transport vesicle approaches the target membrane. (2) The transport vesicle tethers to the target membrane by forming initial physical links between the vesicles and the acceptor membrane through tethering factors. (3) The v-SNARE on the transport vesicle and three t-SNAREs on the target compartment form a 4-α-helix bundle, a *trans*-SNARE complex. The SNARE bundle bridges the two membranes, and is thought to overcome the energy barrier preventing the two membranes from fusing.

Regulatory lipids such as DAG, PtdInsP and sterols are required for coordinating protein-mediated vacuole fusion. Vacuole fusion is blocked by ligands that specifically adsorb DAG, PtdInsP and sterols (Mayer *et al.*, 2000; Kato and Wickner, 2001; Fratti *et al.*, 2004). These studies revealed that DAG, PtdInsP, and sterol regulate priming and docking steps in vacuole fusion. DAG mediates fusion by promoting negative membrane curvature at the fusion site (Allan *et al.*, 1978), whereas PtdIns(4,5)P_2 affects fusion event through actin remodeling (Higgs and Pollard, 2000). *In vitro* studies using a reconstituted yeast vacuole SNARE fusion system revealed that addition of individual regulatory lipids, especially DAG and PtdInsP, strongly enhanced lipid mixing of the donor and the acceptor liposomes, compared to vacuolar lipids such as PtdCho and PtdSer alone (Mima *et al.*, 2008).

3.1. Vesicle Tethering at the Golgi

Tethering factors link Golgi cisternae or capture transport vesicles in the proximity of the cisternae prior to fusion. Many tethers interact with Rab GTPase that cycle between the cytosol and the membrane (Stenmark, 2009). GDP-bound Rab is complexed with guanine nucleotide dissociation inhibitor (GDI) and recruited to membranes via the geranyl-geranyl group with the aid of a GDI displacement factor. Multimeric tethers interact with and activate GDP-bound Rab1 through intrinsic guanine nucleotide exchange activity (Jones et al., 2000b; Wang et al., 2000a). Four multi-subunit tethers are known to be involved in tethering of vesicles at the Golgi; transport protein particle (TRAPP) I (ER-Golgi), TRAPP II (intra-Golgi/endosome-late Golgi), conserved oligomeric Golgi (COG) (endosome-early Golgi) and Golgi-associated retrograde protein (GARP) (endosome-late Golgi) (Cai et al., 2007). In addition, multimeric tethers interact with coat proteins; mutations in TRAPP II, which interacts with COPI, increases COPI-coated vesicles near the Golgi and accumulation of cargos in an early Golgi compartment (Yamasaki et al., 2009).

On the other hand, coiled-coil tethers are monomeric proteins that have a long rod-like structure for linking two membranes. Amongst more than twenty Golgi-localized coiled-coil tethers, several are shown to be involved in vesicle tethering to the Golgi such as p115, GM130, giantin, golgin84, GMAP210 (ER-Golgi and intra-Golgi traffic), golgin97, and GCC185 (endosome-TGN) (Lupashin and Sztul, 2005, Reddy et al., 2006; Drin et al., 2008). In contrast to multimeric tethers that bind GDP-bound Rabs, P115 is an effector of Rab1 that binds to the GTP-bound form but not the GDP-bound form (Allan et al., 2000). GDP-restricted mutant Rab1 blocked fusion of COPII vesicle with the Golgi membrane. Structural studies suggested a model for membrane tethering followed by membrane fusion (An et al., 2009). p115 is initially recruited to the donor membrane through interaction with GTP-bound Rab1. As p115 interacts with the long coiled-coil tether giantin, the donor membrane is attached to the acceptor membrane. Subsequently, p115 is handed to the more compact tether GM130, resulting in a physical draw between the two membranes. Finally GM130 binds the syntaxin 5 t-SNARE at the acceptor membrane in a p115-regulated manner and SNARE assembly drives membrane fusion (Diao et al., 2008). GMAP210 contains a curvature-sensing ALPS motif at the N-terminus and a GRIP-related Arf binding domain at the C-terminus that interacts with a GTP-bound Arf1 recruited on the Golgi membrane, thus mediating tethering of highly curved vesicular membranes with the flat Golgi surface (Drin et al., 2008).

3.2. Vesicle Fusion Mediated by SNAREs

SNARE proteins are necessary for membrane fusion in all eukaryotic cells. SNARE proteins are diverse and share approximately 20~30% identity, but contain a characteristic ~70 residue SNARE motif with heptad repeats (Kloepper et al., 2007). SNARE motifs are classified into four types (R-, Qa-, Qb, Qc) that are present in v-SNAREs on transport intermediates and t-SNAREs on the target compartment. t-SNAREs are further classified into two syntaxin- or SNAP-25-types (Zhao et al., 2007). v-SNAREs and syntaxin-type t-SNAREs are type II integral membrane proteins with a single transmembrane domain. SNAP-25-type tSNAREs lack the transmembrane domain and are anchored to the membrane through thioester-linked acyl groups. Crystal structure analysis of SNARE complexes showed that SNARE motifs assemble into parallel, twisted, coiled-coil, four-helix bundles by burying the hydrophobic residues inside the core. One of the helices is contributed by v-SNARE, with the other three helices provided from t-SNAREs; one from syntaxin and two from SNAP-25.

Two models have been proposed for SNARE-mediated membrane fusion (Jahn and Scheller, 2006). In the first, SNARE assembly exerts a mechanical force on membranes that causes membrane fusion. The linkers between the helix bundles and the transmembrane domains are rigid, thus a tight four-helix bundle presses and deforms to drive hemifusion of the two bilayers. Single molecule force measurements using atomic force microscope/spectroscopy revealed that the trans-interaction of the SNARE helix bundles provides the necessary pulling force to destabilize the lipid bilayers and facilitate hemifusion (Abdulreda *et al.*, 2009). The second model suggests that putative amphiphilic regions of the SNAREs perturbs the hydrophilic-hydrophobic boundary and leads to formation of non-bilayer states without the use of mechanical force. The conserved H3 domain of SNAP-25 has been predicted to form an α-helix that contains a repeating pattern of hydrophobic amino acids characteristic of amphipathic α-helices (Zhong *et al.*, 1997). Whether additional fusogenic proteins are involved in SNARE-mediated membrane fusion, and whether lipid composition changes at the site of SNARE-mediated membrane fusion, remains to be elucidated.

4. DISEASES RELATED TO LIPID METABOLISM AND GOLGI FUNCTION

4.1. Lowe Syndrome

Lowe syndrome is a rare X-linked disorder that is characterized by congenital cataracts, cognitive defects and renal ion transport insufficiency (Olivos-Glander *et al.*, 1995). Kidney proximal tubule cells from Lowe syndrome patients have elevated levels of PtdIns(4,5)P_2 in the Golgi apparatus and PM (Olivos-Glander *et al.*, 1995). OCRL1 was identified as the gene responsible for this disorder, and missense mutations were found clustered in the 5-phosphatase domain. Consistent with the involvement of PtdIns(4,5)P_2 in actin assembly through effector proteins (Insall and Weiner, 2001), fibroblasts from Lowe syndrome patients have a defective actin cytoskeleton that is necessary for vesicle trafficking from the TGN (Suchy and Nussbaum, 2002). Indeed, Lowe syndrome patients have abnormally high levels of serum lysosomal hydrolases (Ungewickell and Majerus, 1999), probably due to defective formation of clathrin-coated vesicles mediating transport of lysosomal enzymes from the Golgi to endosomes (Choudhury *et al.*, 2005). As OCRL1 depletion leads to PtdIns(4)P reduction within the Golgi (Choudhury *et al.*, 2005), the role of OCRL1 might be production of PtdIns(4)P that mediates recruitment of clathrin adaptors and EpsinR (Section 2.1., 2.3.2.).

4.2. Congenital Disorders of Glycosylation

Autosomal recessive congenital disorders of glycosylation (CDG) are characterized by defects in N-linked glycan biogenesis resulting from mutations in genes encoding proteins in the glycosylation pathway (Marquardt and Denecke, 2003). Since glycosylation occurs in all cells, CDG patients show multi-system abnormalities such as cognitive impairment, seizures, hypotonia, liver malfunctions, coagulopathy, and dysmorphia. Approximately 25% of CDG patients die from severe infections or multiple organ failure. CDG are classified into Type I defects due to decreased synthesis of the dolichol-linked precursors of N-linked glycosylation in the ER, and Type II defects due to impairment of subsequent trimming of the protein-bound oligosaccharide and addition of terminal sugars in the Golgi apparatus. Although defects in enzymes or transporter directly involved in glycosylation are often found in CDG patients, CDG-IIe, a sub-class of Type II, is caused by a mutation in genes that encode components of the COG tethering complex (Wu *et al.*, 2004; Foulquier, 2009). COG has been shown to be involved in controlling stability of variety of Golgi-associated glycosylation enzymes as well as in retrograde membrane trafficking of Golgi resident proteins and membrane tethering. Golgi glycosylation enzymes including galactosyltransferases are mislocalized and/or rapidly degraded in COG-deficient cells (Wu *et al.*, 2004).

4.3. Tangier Disease

Tangier disease (TD) is a rare autosomal recessive disorder characterized by almost complete absence of plasma HDL, and reduced efflux of cholesterol, especially from macrophages and other reticulo-endothelial cells, to the extracellular acceptor apoAI (Bodzioch *et al.*, 1999). This leads to cholesterol ester accumulation in these cells and is associated with increased susceptibility to atherosclerosis. Mutations in the ABCA1 gene have been identified in TD patients. Cells from mouse ABCA1 knockouts and TD patients showed an extensively expanded Golgi complex, probably due to severe inhibition of lipid export from the Golgi to the PM (Orso *et al.*, 2000). ABCA1 expression at the PM surface is controlled in response to cellular cholesterol levels and mediates export of excess cholesterol (Langmann *et al.*, 1999). However, ABCA1 also localizes in intracellular compartments together with apoA-I (Orso *et al.*, 2000; Smith *et al.*, 2002). A model was proposed that ABCA1 shuttles from the PM to intracellular compartments (including the Golgi apparatus) along with apoA-I, and extracts cholesterol and phospholipids into the vesicle for association with apoA-I, which is subsequently released upon PM fusion. Thus, ABCA1 might mediate extraction and extrusion of excess cholesterol by direct interactions with cellular compartments.

5. CONCLUSIONS AND FUTURE DIRECTIONS

Pathways reported above indicate that Golgi dynamics require the coordinated activities of both lipid and protein machineries. Individual pathways are linked by common regulatory lipids and proteins to generate integrated networks. In addition to identifying the missing effectors and regulators within these pathways, future challenges in this area are to elucidate how lipid rafts and cytoskeleton organizations are integrated with Golgi function. In addition, it will be very interesting to investigate whether this process is altered by oxidative stress or under pathological conditions that occur in stroke and cardiovascular disease.

REFERENCES

Abdulreda, M.H., Bhalla, A., Rico, F., Berggren, P.O., Chapman, E.R., and Moy, V.T. (2009). Pulling force generated by interacting SNAREs facilitates membrane hemifusion. *Integr Biol (*Camb) *1*, 301-310.

Adams, C.M., Reitz, J., De Brabander, J.K., Feramisco, J.D., Li, L., Brown, M.S., and Goldstein, J.L. (2004). Cholesterol and 25-hydroxycholesterol inhibit activation of SREBPs by different mechanisms, both involving SCAP and Insigs. *J Biol Chem 279*, 52772-52780.

Aikawa, Y., Kuraoka, A., Kondo, H., Kawabuchi, M., and Watanabe, T. (1999). Involvement of PITPnm, a mammalian homologue of Drosophila rdgB, in phosphoinositide synthesis on Golgi membranes. *J Biol Chem 274*, 20569-20577.

Allan, B.B., Moyer, B.D., and Balch, W.E. (2000). Rab1 recruitment of p115 into a cis-SNARE complex: programming budding COPII vesicles for fusion. *Science 289*, 444-448.

Allan, D., Thomas, P., and Michell, R.H. (1978). Rapid transbilayer diffusion of 1,2-diacylglycerol and its relevance to control of membrane curvature. *Nature 276*, 289-290.

Allen-Baume, V., Segui, B., and Cockcroft, S. (2002). Current thoughts on the phosphatidylinositol transfer protein family. *FEBS Lett 531*, 74-80.

Alphey, L., Jimenez, J., and Glover, D. (1998). A Drosophila homologue of oxysterol binding protein (OSBP)--implications for the role of OSBP. *Biochim Biophys Acta 1395*, 159-164.

Altmann, S.W., Davis, H.R., Jr., Zhu, L.J., Yao, X., Hoos, L.M., Tetzloff, G., Iyer, S.P., Maguire, M., Golovko, A., Zeng, M., Wang, L., Murgolo, N., and Graziano, M.P. (2004). Niemann-Pick C1 Like 1 protein is critical for intestinal cholesterol absorption. *Science 303*, 1201-1204.

An, Y., Chen, C.Y., Moyer, B., Rotkiewicz, P., Elsliger, M.A., Godzik, A., Wilson, I.A., and Balch, W.E. (2009). Structural and functional analysis of the globular head domain of p115 provides insight into membrane tethering. *J Mol Biol 391*, 26-41.

Antonny, B. (2006). Membrane deformation by protein coats. *Curr Opin Cell Biol 18*, 386-394.

Antonny, B., Huber, I., Paris, S., Chabre, M., and Cassel, D. (1997). Activation of ADP-ribosylation factor 1 GTPase-activating protein by phosphatidylcholine-derived diacylglycerols. *J Biol Chem 272*, 30848-30851.

Asp, L., Kartberg, F., Fernandez-Rodriguez, J., Smedh, M., Elsner, M., Laporte, F., Barcena, M., Jansen, K.A., Valentijn, J.A., Koster, A.J., Bergeron, J.J., and Nilsson, T. (2009). Early stages of Golgi vesicle and tubule formation require diacylglycerol. *Mol Biol Cell 20*, 780-790.

Balla, A., and Balla, T. (2006). Phosphatidylinositol 4-kinases: old enzymes with emerging functions. *Trends Cell Biol 16*, 351-361.

Balla, A., Tuymetova, G., Tsiomenko, A., Varnai, P., and Balla, T. (2005). A plasma membrane pool of phosphatidylinositol 4-phosphate is generated by phosphatidylinositol 4-kinase type-III alpha: studies with the PH domains of the oxysterol binding protein and FAPP1. *Mol Biol Cell 16*, 1282-1295.

Banerji, S., Ngo, M., Lane, C., Robinson, C.A., Minogue, S. and Ridgway, N.D. (2010) Oxysterol binding protein-dependent activation of sphingomyelin synthesis in the Golgi apparatus requires PtdIns 4-kinase IIα. *Mol. Biol. Cell. In press.*

Bard, F., and Malhotra, V. (2006). The formation of TGN-to-plasma-membrane transport carriers. *Annu Rev Cell Dev Biol 22*, 439-455.

Baron, C.L., and Malhotra, V. (2002). Role of diacylglycerol in PKD recruitment to the TGN and protein transport to the plasma membrane. *Science 295*, 325-328.

Bechler, M.E., Doody, A.M., Racoosin, E., Lin, L., Lee, K.H., and Brown, W.J. (2010). The phospholipase complex PAFAH Ib regulates the functional organization of the Golgi complex. *J Cell Biol 190*, 45-53.

Beck, R., Rawet, M., Wieland, F.T., and Cassel, D. (2009). The COPI system: molecular mechanisms and function. *FEBS Lett 583*, 2701-2709.

Beck, R., Sun, Z., Adolf, F., Rutz, C., Bassler, J., Wild, K., Sinning, I., Hurt, E., Brugger, B., Bethune, J., and Wieland, F. (2008). *Membrane curvature induced by Arf1-GTP is essential for vesicle formation.* Proc Natl Acad Sci U S A *105*, 11731-11736.

Bennett, M.K., Lopez, J.M., Sanchez, H.B., and Osborne, T.F. (1995). Sterol regulation of fatty acid synthase promoter. Coordinate feedback regulation of two major lipid pathways. *J Biol Chem 270*, 25578-25583.

Bi, X., Corpina, R.A., and Goldberg, J. (2002). Structure of the Sec23/24-Sar1 pre-budding complex of the COPII vesicle coat. *Nature 419*, 271-277.

Bielli, A., Haney, C.J., Gabreski, G., Watkins, S.C., Bannykh, S.I., and Aridor, M. (2005). Regulation of Sar1 NH2 terminus by GTP binding and hydrolysis promotes membrane deformation to control COPII vesicle fission. *J Cell Biol 171*, 919-924.

Bigay, J., Casella, J.F., Drin, G., Mesmin, B., and Antonny, B. (2005). ArfGAP1 responds to membrane curvature through the folding of a lipid packing sensor motif. *EMBO J 24*, 2244-2253.

Bigay, J., Gounon, P., Robineau, S., and Antonny, B. (2003). Lipid packing sensed by ArfGAP1 couples COPI coat disassembly to membrane bilayer curvature. *Nature 426*, 563-566.

Billah, M.M., Eckel, S., Mullmann, T.J., Egan, R.W., and Siegel, M.I. (1989). Phosphatidylcholine hydrolysis by phospholipase D determines phosphatidate and diglyceride levels in chemotactic peptide-stimulated human neutrophils. Involvement of phosphatidate phosphohydrolase in signal transduction. *J Biol Chem 264*, 17069-17077.

Blagoveshchenskaya, A., Cheong, F.Y., Rohde, H.M., Glover, G., Knodler, A., Nicolson, T., Boehmelt, G., and Mayinger, P. (2008). Integration of Golgi trafficking and growth factor signaling by the lipid phosphatase SAC1. *J Cell Biol 180*, 803-812.

Bodzioch, M., Orso, E., Klucken, J., Langmann, T., Bottcher, A., Diederich, W., Drobnik, W., Barlage, S., Buchler, C., Porsch-Ozcurumez, M., Kaminski, W.E., Hahmann, H.W., Oette, K., Rothe, G., Aslanidis, C., Lackner, K.J., and Schmitz, G. (1999). The gene encoding ATP-binding cassette transporter 1 is mutated in Tangier disease. *Nat Genet 22*, 347-351.

Boman, A.L., Zhang, C., Zhu, X., and Kahn, R.A. (2000). A family of ADP-ribosylation factor effectors that can alter membrane transport through the trans-Golgi. *Mol Biol Cell 11*, 1241-1255.

Bowden, K., and Ridgway, N.D. (2008). OSBP negatively regulates ABCA1 protein stability. *J Biol Chem 283*, 18210-18217.

Breathnach, C.S. (2001). Johann Ludwig Wilhelm Thudichum 1829-1901, bane of the Protagonisers. *Hist Psychiatry 12*, 283-296.

Bretscher, M.S., and Munro, S. (1993). Cholesterol and the Golgi apparatus. *Science 261*, 1280-1281.

Brown, A.J., Sun, L., Feramisco, J.D., Brown, M.S., and Goldstein, J.L. (2002). Cholesterol addition to ER membranes alters conformation of SCAP, the SREBP escort protein that regulates cholesterol metabolism. *Mol Cell 10*, 237-245.

Brown, D.A., and London, E. (1998). Functions of lipid rafts in biological membranes. *Annu Rev Cell Dev Biol 14*, 111-136.

Brown, M.S., and Goldstein, J.L. (1997). The SREBP pathway: regulation of cholesterol metabolism by proteolysis of a membrane-bound transcription factor. *Cell 89*, 331-340.

Buton, X., Herve, P., Kubelt, J., Tannert, A., Burger, K.N., Fellmann, P., Muller, P., Herrmann, A., Seigneuret, M., and Devaux, P.F. (2002). Transbilayer movement of monohexosylsphingolipids in endoplasmic reticulum and Golgi membranes. *Biochemistry 41*, 13106-13115.

Cai, H., Reinisch, K., and Ferro-Novick, S. (2007). Coats, tethers, Rabs, and SNAREs work together to mediate the intracellular destination of a transport vesicle. *Dev Cell 12*, 671-682.

Canagarajah, B.J., Hummer, G., Prinz, W.A., and Hurley, J.H. (2008). Dynamics of cholesterol exchange in the oxysterol binding protein family. *J Mol Biol 378*, 737-748.

Cao, X., Coskun, U., Rossle, M., Buschhorn, S.B., Grzybek, M., Dafforn, T.R., Lenoir, M., Overduin, M., and Simons, K. (2009). *Golgi protein FAPP2 tubulates membranes*. Proc Natl Acad Sci U S A *106*, 21121-21125.

Carvou, N., Holic, R., Li, M., Futter, C., Skippen, A., and Cockcroft, S. (2010). Phosphatidylinositol- and phosphatidylcholine-transfer activity of PITPβeta is essential for COPI-mediated retrograde transport from the Golgi to the endoplasmic reticulum. *J Cell Sci 123*, 1262-1273.

Chen, W., Feng, Y., Chen, D., and Wandinger-Ness, A. (1998). Rab11 is required for trans-golgi network-to-plasma membrane transport and a preferential target for GDP dissociation inhibitor. *Mol Biol Cell 9*, 3241-3257.

Chernomordik, L., Chanturiya, A., Green, J., and Zimmerberg, J. (1995). The hemifusion intermediate and its conversion to complete fusion: regulation by membrane composition. *Biophys J 69*, 922-929.

Chernomordik, L.V., and Kozlov, M.M. (2008). Mechanics of membrane fusion. *Nat Struct Mol Biol 15*, 675-683.

Cheruku, S.R., Xu, Z., Dutia, R., Lobel, P., and Storch, J. (2006). Mechanism of cholesterol transfer from the Niemann-Pick type C2 protein to model membranes supports a role in lysosomal cholesterol transport. *J Biol Chem 281*, 31594-31604.

Choudhury, R., Diao, A., Zhang, F., Eisenberg, E., Saint-Pol, A., Williams, C., Konstantakopoulos, A., Lucocq, J., Johannes, L., Rabouille, C., Greene, L.E., and Lowe, M. (2005). Lowe syndrome protein OCRL1 interacts with clathrin and regulates protein trafficking between endosomes and the trans-Golgi network. *Mol Biol Cell 16*, 3467-3479.

Cockcroft, S., and Carvou, N. (2007). Biochemical and biological functions of class I phosphatidylinositol transfer proteins. *Biochim Biophys Acta 1771*, 677-691.

Collins, B.M., Praefcke, G.J., Robinson, M.S., and Owen, D.J. (2003). Structural basis for binding of accessory proteins by the appendage domain of GGAs. *Nat Struct Biol 10*, 607-613.

Collins, R.N., and Warren, G. (1992). Sphingolipid transport in mitotic HeLa cells. *J Biol Chem 267*, 24906-24911.

Corda, D., Hidalgo Carcedo, C., Bonazzi, M., Luini, A., and Spano, S. (2002). Molecular aspects of membrane fission in the secretory pathway. *Cell Mol Life Sci 59*, 1819-1832.

Coskun, U., and Simons, K. (2010). Membrane rafting: from apical sorting to phase segregation. *FEBS Lett 584*, 1685-1693.

Coste, H., Martel, M.B., and Got, R. (1986). Topology of glucosylceramide synthesis in Golgi membranes from porcine submaxillary glands. *Biochim Biophys Acta 858*, 6-12.

Cukierman, E., Huber, I., Rotman, M., and Cassel, D. (1995). The ARF1 GTPase-activating protein: zinc finger motif and Golgi complex localization. *Science 270*, 1999-2002.

D'Angelo, G., Polishchuk, E., Di Tullio, G., Santoro, M., Di Campli, A., Godi, A., West, G., Bielawski, J., Chuang, C.C., van der Spoel, A.C., Platt, F.M., Hannun, Y.A., Polishchuk, R., Mattjus, P., and De Matteis, M.A. (2007). Glycosphingolipid synthesis requires FAPP2 transfer of glucosylceramide. *Nature 449*, 62-67.

Das, A.K., Horie, S., and Hajra, A.K. (1992). Biosynthesis of glycerolipid precursors in rat liver peroxisomes and their transport and conversion to phosphatidate in the endoplasmic reticulum. *J Biol Chem 267*, 9724-9730.

Dascher, C., and Balch, W.E. (1994). Dominant inhibitory mutants of ARF1 block endoplasmic reticulum to Golgi transport and trigger disassembly of the Golgi apparatus. *J Biol Chem 269*, 1437-1448.

de Barry, J., Janoshazi, A., Dupont, J.L., Procksch, O., Chasserot-Golaz, S., Jeromin, A., and Vitale, N. (2006). Functional implication of neuronal calcium sensor-1 and phosphoinositol 4-kinase-beta interaction in regulated exocytosis of PC12 cells. *J Biol Chem 281*, 18098-18111.

de Figueiredo, P., Drecktrah, D., Katzenellenbogen, J.A., Strang, M., and Brown, W.J. (1998). *Evidence that phospholipase A2 activity is required for Golgi complex and trans Golgi network membrane tubulation.* Proc Natl Acad Sci U S A *95*, 8642-8647.

de Figueiredo, P., Polizotto, R.S., Drecktrah, D., and Brown, W.J. (1999). Membrane tubule-mediated reassembly and maintenance of the Golgi complex is disrupted by phospholipase A2 antagonists. *Mol Biol Cell 10*, 1763-1782.

de Graaf, P., Zwart, W.T., van Dijken, R.A., Deneka, M., Schulz, T.K., Geijsen, N., Coffer, P.J., Gadella, B.M., Verkleij, A.J., van der Sluijs, P., and van Bergen en Henegouwen, P.M. (2004). Phosphatidylinositol 4-kinasebeta is critical for functional association of rab11 with the Golgi complex. *Mol Biol Cell 15*, 2038-2047.

De Matteis, M.A., Di Campli, A., and Godi, A. (2005). The role of the phosphoinositides at the Golgi complex. *Biochim Biophys Acta 1744*, 396-405.

Dell'Angelica, E.C., Puertollano, R., Mullins, C., Aguilar, R.C., Vargas, J.D., Hartnell, L.M., and Bonifacino, J.S. (2000). GGAs: a family of ADP ribosylation factor-binding proteins related to adaptors and associated with the Golgi complex. *J Cell Biol 149*, 81-94.

Deng, Y., Golinelli-Cohen, M.P., Smirnova, E., and Jackson, C.L. (2009). A COPI coat subunit interacts directly with an early-Golgi localized Arf exchange factor. *EMBO Rep 10*, 58-64.

Diao, A., Frost, L., Morohashi, Y., and Lowe, M. (2008). Coordination of golgin tethering and SNARE assembly: GM130 binds syntaxin 5 in a p115-regulated manner. *J Biol Chem 283*, 6957-6967.

Diaz Anel, A.M., and Malhotra, V. (2005). PKCeta is required for beta1gamma2/beta3gamma2- and PKD-mediated transport to the cell surface and the organization of the Golgi apparatus. *J Cell Biol 169*, 83-91.

Ding, T., Li, Z., Hailemariam, T., Mukherjee, S., Maxfield, F.R., Wu, M.P., and Jiang, X.C. (2008). SMS overexpression and knockdown: impact on cellular sphingomyelin and diacylglycerol metabolism, and cell apoptosis. *J Lipid Res 49*, 376-385.

Doray, B., Ghosh, P., Griffith, J., Geuze, H.J., and Kornfeld, S. (2002). Cooperation of GGAs and AP-1 in packaging MPRs at the trans-Golgi network. *Science 297*, 1700-1703.

Drecktrah, D., Chambers, K., Racoosin, E.L., Cluett, E.B., Gucwa, A., Jackson, B., and Brown, W.J. (2003). Inhibition of a Golgi complex lysophospholipid acyltransferase induces membrane tubule formation and retrograde trafficking. *Mol Biol Cell 14*, 3459-3469.

Dressman, M.A., Olivos-Glander, I.M., Nussbaum, R.L., and Suchy, S.F. (2000). Ocrl1, a PtdIns(4,5)P(2) 5-phosphatase, is localized to the trans-Golgi network of fibroblasts and epithelial cells. *J Histochem Cytochem 48*, 179-190.

Drin, G., Casella, J.F., Gautier, R., Boehmer, T., Schwartz, T.U., and Antonny, B. (2007). A general amphipathic alpha-helical motif for sensing membrane curvature. *Nat Struct Mol Biol 14*, 138-146.

Drin, G., Morello, V., Casella, J.F., Gounon, P., and Antonny, B. (2008). Asymmetric tethering of flat and curved lipid membranes by a golgin. *Science 320*, 670-673.

Dumaresq-Doiron, K., Savard, M.F., Akam, S., Costantino, S., and Lefrancois, S. (2010). The phosphatidylinositol 4-kinase PI4KIIIα is required for the recruitment of GBF1 to Golgi membranes. *J Cell Sci 123*, 2273-2280.

Eastman, S.J., Hope, M.J., and Cullis, P.R. (1991). Transbilayer transport of phosphatidic acid in response to transmembrane pH gradients. *Biochemistry 30*, 1740-1745.

Eckford, P.D., and Sharom, F.J. (2005). The reconstituted P-glycoprotein multidrug transporter is a flippase for glucosylceramide and other simple glycosphingolipids. *Biochem J 389*, 517-526.

Erdmann, K.S., Mao, Y., McCrea, H.J., Zoncu, R., Lee, S., Paradise, S., Modregger, J., Biemesderfer, D., Toomre, D., and De Camilli, P. (2007). A role of the Lowe syndrome protein OCRL in early steps of the endocytic pathway. *Dev Cell 13*, 377-390.

Ericsson, J., Jackson, S.M., Kim, J.B., Spiegelman, B.M., and Edwards, P.A. (1997). Identification of glycerol-3-phosphate acyltransferase as an adipocyte determination and differentiation factor 1- and sterol regulatory element-binding protein-responsive gene. *J Biol Chem 272*, 7298-7305.

Eugster, A., Frigerio, G., Dale, M., and Duden, R. (2000). COP I domains required for coatomer integrity, and novel interactions with ARF and ARF-GAP. *EMBO J 19*, 3905-3917.

Farsad, K., and De Camilli, P. (2003). Mechanisms of membrane deformation. *Curr Opin Cell Biol 15*, 372-381.

Faucherre, A., Desbois, P., Satre, V., Lunardi, J., Dorseuil, O., and Gacon, G. (2003). Lowe syndrome protein OCRL1 interacts with Rac GTPase in the trans-Golgi network. *Hum Mol Genet 12*, 2449-2456.

Fernandez-Ulibarri, I., Vilella, M., Lazaro-Dieguez, F., Sarri, E., Martinez, S.E., Jimenez, N., Claro, E., Merida, I., Burger, K.N., and Egea, G. (2007). Diacylglycerol is required for the formation of COPI vesicles in the Golgi-to-ER transport pathway. *Mol Biol Cell 18*, 3250-3263.

Folsch, H., Pypaert, M., Maday, S., Pelletier, L., and Mellman, I. (2003). The AP-1A and AP-1B clathrin adaptor complexes define biochemically and functionally distinct membrane domains. *J Cell Biol 163*, 351-362.

Ford, M.G., Mills, I.G., Peter, B.J., Vallis, Y., Praefcke, G.J., Evans, P.R., and McMahon, H.T. (2002). Curvature of clathrin-coated pits driven by epsin. *Nature 419*, 361-366.

Foulquier, F. (2009). COG defects, birth and rise! *Biochim Biophys Acta 1792*, 896-902.

Fratti, R.A., Jun, Y., Merz, A.J., Margolis, N., and Wickner, W. (2004). Interdependent assembly of specific regulatory lipids and membrane fusion proteins into the vertex ring domain of docked vacuoles. *J Cell Biol 167*, 1087-1098.

Freyberg, Z., Bourgoin, S., and Shields, D. (2002). Phospholipase D2 is localized to the rims of the Golgi apparatus in mammalian cells. *Mol Biol Cell 13*, 3930-3942.

Furuita, K., Jee, J., Fukada, H., Mishima, M., and Kojima, C. (2010). Electrostatic interaction between oxysterol-binding protein and VAMP-associated protein A revealed by NMR and mutagenesis studies. *J Biol Chem 285*, 12961-12970.

Futerman, A.H., and Pagano, R.E. (1991). Determination of the intracellular sites and topology of glucosylceramide synthesis in rat liver. Biochem J *280 (Pt 2)*, 295-302.

Ghosh, P., Griffith, J., Geuze, H.J., and Kornfeld, S. (2003). Mammalian GGAs act together to sort mannose 6-phosphate receptors. *J Cell Biol 163*, 755-766.

Glick, B.S., and Nakano, A. (2009). Membrane traffic within the Golgi apparatus. *Annu Rev Cell Dev Biol 25*, 113-132.

Godi, A., Di Campli, A., Konstantakopoulos, A., Di Tullio, G., Alessi, D.R., Kular, G.S., Daniele, T., Marra, P., Lucocq, J.M., and De Matteis, M.A. (2004). FAPPs control Golgi-to-cell-surface membrane traffic by binding to ARF and PtdIns(4)P. *Nat Cell Biol 6*, 393-404.

Godi, A., Pertile, P., Meyers, R., Marra, P., Di Tullio, G., Iurisci, C., Luini, A., Corda, D., and De Matteis, M.A. (1999). ARF mediates recruitment of PtdIns-4-OH kinase-beta and stimulates synthesis of PtdIns(4,5)P2 on the Golgi complex. *Nat Cell Biol 1*, 280-287.

Goldstein, J.L., DeBose-Boyd, R.A., and Brown, M.S. (2006a). Protein sensors for membrane sterols. *Cell 124*, 35-46.

Goldstein, J.L., Glossip, D., Nayak, S., and Kornfeld, K. (2006b). The CRAL/TRIO and GOLD domain protein CGR-1 promotes induction of vulval cell fates in Caenorhabditis elegans and interacts genetically with the Ras signaling pathway. *Genetics 172*, 929-942.

Hakomori, S. (2000). Traveling for the glycosphingolipid path. *Glycoconj J 17*, 627-647.

Halter, D., Neumann, S., van Dijk, S.M., Wolthoorn, J., de Maziere, A.M., Vieira, O.V., Mattjus, P., Klumperman, J., van Meer, G., and Sprong, H. (2007). Pre- and post-Golgi translocation of glucosylceramide in glycosphingolipid synthesis. *J Cell Biol 179*, 101-115.

Hanada, K., Kumagai, K., Tomishige, N., and Yamaji, T. (2009). CERT-mediated trafficking of ceramide. *Biochim Biophys Acta 1791*, 684-691.

Hanada, K., Kumagai, K., Yasuda, S., Miura, Y., Kawano, M., Fukasawa, M., and Nishijima, M. (2003). Molecular machinery for non-vesicular trafficking of ceramide. *Nature 426*, 803-809.

Hausser, A., Link, G., Hoene, M., Russo, C., Selchow, O., and Pfizenmaier, K. (2006). Phospho-specific binding of 14-3-3 proteins to phosphatidylinositol 4-kinase III beta protects from dephosphorylation and stabilizes lipid kinase activity. *J Cell Sci 119*, 3613-3621.

Hausser, A., Storz, P., Martens, S., Link, G., Toker, A., and Pfizenmaier, K. (2005). Protein kinase D regulates vesicular transport by phosphorylating and activating phosphatidylinositol-4 kinase IIIbeta at the Golgi complex. *Nat Cell Biol 7*, 880-886.

Haynes, L.P., Thomas, G.M., and Burgoyne, R.D. (2005). Interaction of neuronal calcium sensor-1 and ADP-ribosylation factor 1 allows bidirectional control of phosphatidylinositol 4-kinase beta and trans-Golgi network-plasma membrane traffic. *J Biol Chem 280*, 6047-6054.

Heino, S., Lusa, S., Somerharju, P., Ehnholm, C., Olkkonen, V.M., and Ikonen, E. (2000). *Dissecting the role of the golgi complex and lipid rafts in biosynthetic transport of cholesterol to the cell surface*. Proc Natl Acad Sci U S A *97*, 8375-8380.

Heldwein, E.E., Macia, E., Wang, J., Yin, H.L., Kirchhausen, T., and Harrison, S.C. (2004). *Crystal structure of the clathrin adaptor protein 1 core*. Proc Natl Acad Sci U S A *101*, 14108-14113.

Higgs, H.N., and Pollard, T.D. (2000). Activation by Cdc42 and PIP(2) of Wiskott-Aldrich syndrome protein (WASp) stimulates actin nucleation by Arp2/3 complex. *J Cell Biol 150*, 1311-1320.

Hirst, J., Lui, W.W., Bright, N.A., Totty, N., Seaman, M.N., and Robinson, M.S. (2000). A family of proteins with gamma-adaptin and VHS domains that facilitate trafficking between the trans-Golgi network and the vacuole/lysosome. *J Cell Biol 149*, 67-80.

Hirst, J., Miller, S.E., Taylor, M.J., von Mollard, G.F., and Robinson, M.S. (2004). EpsinR is an adaptor for the SNARE protein Vti1b. *Mol Biol Cell 15*, 5593-5602.

Hirst, J., Motley, A., Harasaki, K., Peak Chew, S.Y., and Robinson, M.S. (2003). EpsinR: an ENTH domain-containing protein that interacts with AP-1. *Mol Biol Cell 14*, 625-641.

Hooper, N.M. (1999). Detergent-insoluble glycosphingolipid/cholesterol-rich membrane domains, lipid rafts and caveolae (review). *Mol Membr Biol 16*, 145-156.

Horvath, C.A., Vanden Broeck, D., Boulet, G.A., Bogers, J., and De Wolf, M.J. (2007). Epsin: inducing membrane curvature. *Int J Biochem Cell Biol 39*, 1765-1770.

Hughes, H., and Stephens, D.J. (2008). Assembly, organization, and function of the COPII coat. *Histochem Cell Biol 129*, 129-151.

Huitema, K., van den Dikkenberg, J., Brouwers, J.F., and Holthuis, J.C. (2004). Identification of a family of animal sphingomyelin synthases. *EMBO J 23*, 33-44.

Hyman, J., Chen, H., Di Fiore, P.P., De Camilli, P., and Brunger, A.T. (2000). Epsin 1 undergoes nucleocytosolic shuttling and its eps15 interactor NH(2)-terminal homology

(ENTH) domain, structurally similar to Armadillo and HEAT repeats, interacts with the transcription factor promyelocytic leukemia Zn(2)+ finger protein (PLZF). *J Cell Biol 149*, 537-546.

Hyvola, N., Diao, A., McKenzie, E., Skippen, A., Cockcroft, S., and Lowe, M. (2006). Membrane targeting and activation of the Lowe syndrome protein OCRL1 by rab GTPases. *EMBO J 25*, 3750-3761.

Im, Y.J., Raychaudhuri, S., Prinz, W.A., and Hurley, J.H. (2005) Structural mechanims for sterol sensing and transport by OSBP-related proteins. *Nature 437,* 154-158.

Insall, R.H., and Weiner, O.D. (2001). PIP3, PIP2, and cell movement--similar messages, different meanings? *Dev Cell 1*, 743-747.

Jacob, T.C., and Kaplan, J.M. (2003). The EGL-21 carboxypeptidase E facilitates acetylcholine release at Caenorhabditis elegans neuromuscular junctions. *J Neurosci 23*, 2122-2130.

Jahn, R., and Scheller, R.H. (2006). SNAREs--engines for membrane fusion. *Nat Rev Mol Cell Biol 7*, 631-643.

Jeckel, D., Karrenbauer, A., Burger, K.N., van Meer, G., and Wieland, F. (1992). Glucosylceramide is synthesized at the cytosolic surface of various Golgi subfractions. *J Cell Biol 117*, 259-267.

Jefferson, A.B., and Majerus, P.W. (1995). Properties of type II inositol polyphosphate 5-phosphatase. *J Biol Chem 270*, 9370-9377.

Jenkins, G.M., and Frohman, M.A. (2005). Phospholipase D: a lipid centric review. *Cell Mol Life Sci 62*, 2305-2316.

Jiang, B., Brown, J.L., Sheraton, J., Fortin, N., and Bussey, H. (1994). A new family of yeast genes implicated in ergosterol synthesis is related to the human oxysterol binding protein. *Yeast 10*, 341-353.

Jones, D.H., Morris, J.B., Morgan, C.P., Kondo, H., Irvine, R.F., and Cockcroft, S. (2000a). Type I phosphatidylinositol 4-phosphate 5-kinase directly interacts with ADP-ribosylation factor 1 and is responsible for phosphatidylinositol 4,5-bisphosphate synthesis in the golgi compartment. *J Biol Chem 275*, 13962-13966.

Jones, S., Newman, C., Liu, F., and Segev, N. (2000b). The TRAPP complex is a nucleotide exchanger for Ypt1 and Ypt31/32. *Mol Biol Cell 11*, 4403-4411.

Joyce, C.W., Shelness, G.S., Davis, M.A., Lee, R.G., Skinner, K., Anderson, R.A., and Rudel, L.L. (2000). ACAT1 and ACAT2 membrane topology segregates a serine residue essential for activity to opposite sides of the endoplasmic reticulum membrane. *Mol Biol Cell 11*, 3675-3687.

Kai, M., Wada, I., Imai, S., Sakane, F., and Kanoh, H. (1997). Cloning and characterization of two human isozymes of Mg2+-independent phosphatidic acid phosphatase. *J Biol Chem 272*, 24572-24578.

Kaiser, S.E., Brickner, J.H., Reilein, A.R., Fenn, T.D., Walter, P., and Brunger, A.T. (2005). Structural basis of FFAT motif-mediated ER targeting. *Structure 13*, 1035-1045.

Kandutsch, A.A., and Shown, E.P. (1981). Assay of oxysterol-binding protein in a mouse fibroblast, cell-free system. Dissociation constant and other properties of the system. *J Biol Chem 256*, 13068-13073.

Kato, M., and Wickner, W. (2001). Ergosterol is required for the Sec18/ATP-dependent priming step of homotypic vacuole fusion. *EMBO J 20*, 4035-4040.

Kawano, M., Kumagai, K., Nishijima, M., and Hanada, K. (2006). Efficient trafficking of ceramide from the endoplasmic reticulum to the Golgi apparatus requires a VAMP-associated protein-interacting FFAT motif of CERT. *J Biol Chem 281*, 30279-30288.

Kawasaki, M., Shiba, T., Shiba, Y., Yamaguchi, Y., Matsugaki, N., Igarashi, N., Suzuki, M., Kato, R., Kato, K., Nakayama, K., and Wakatsuki, S. (2005). Molecular mechanism of ubiquitin recognition by GGA3 GAT domain. *Genes Cells 10*, 639-654.

Kent, H.M., McMahon, H.T., Evans, P.R., Benmerah, A., and Owen, D.J. (2002). Gamma-adaptin appendage domain: structure and binding site for Eps15 and gamma-synergin. *Structure 10*, 1139-1148.

Kim, Y., Kim, J.E., Lee, S.D., Lee, T.G., Kim, J.H., Park, J.B., Han, J.M., Jang, S.K., Suh, P.G., and Ryu, S.H. (1999). Phospholipase D1 is located and activated by protein kinase C alpha in the plasma membrane in 3Y1 fibroblast cell. *Biochim Biophys Acta 1436*, 319-330.

Klappe, K., Hummel, I., Hoekstra, D., and Kok, J.W. (2009). Lipid dependence of ABC transporter localization and function. *Chem Phys Lipids 161*, 57-64.

Kliouchnikov, L., Bigay, J., Mesmin, B., Parnis, A., Rawet, M., Goldfeder, N., Antonny, B., and Cassel, D. (2009). Discrete determinants in ArfGAP2/3 conferring Golgi localization and regulation by the COPI coat. *Mol Biol Cell 20*, 859-869.

Kloepper, T.H., Kienle, C.N., and Fasshauer, D. (2007). An elaborate classification of SNARE proteins sheds light on the conservation of the eukaryotic endomembrane system. *Mol Biol Cell 18*, 3463-3471.

Ko, D.C., Gordon, M.D., Jin, J.Y., and Scott, M.P. (2001). Dynamic movements of organelles containing Niemann-Pick C1 protein: NPC1 involvement in late endocytic events. *Mol Biol Cell 12*, 601-614.

Kooijman, E.E., Chupin, V., de Kruijff, B., and Burger, K.N. (2003). Modulation of membrane curvature by phosphatidic acid and lysophosphatidic acid. *Traffic 4*, 162-174.

Kudo, N., Kumagai, K., Matsubara, R., Kobayashi, S., Hanada, K., Wakatsuki, S., and Kato, R. (2010). Crystal structures of the CERT START domain with inhibitors provide insights into the mechanism of ceramide transfer. *J Mol Biol 396*, 245-251.

Kumagai, K., Kawano, M., Shinkai-Ouchi, F., Nishijima, M., and Hanada, K. (2007). Interorganelle trafficking of ceramide is regulated by phosphorylation-dependent cooperativity between the PH and START domains of CERT. *J Biol Chem 282*, 17758-17766.

Lagace, T.A., Byers, D.M., Cook, H.W., and Ridgway, N.D. (1997). Altered regulation of cholesterol and cholesteryl ester synthesis in Chinese-hamster ovary cells overexpressing the oxysterol-binding protein is dependent on the pleckstrin homology domain. *Biochem J 326 (Pt 1)*, 205-213.

Lagace, T.A., Byers, D.M., Cook, H.W., and Ridgway, N.D. (1999). Chinese hamster ovary cells overexpressing the oxysterol binding protein (OSBP) display enhanced synthesis of sphingomyelin in response to 25-hydroxycholesterol. *J Lipid Res 40*, 109-116.

Lagace, T.A., Storey, M.K., and Ridgway, N.D. (2000). Regulation of phosphatidylcholine metabolism in Chinese hamster ovary cells by the sterol regulatory element-binding protein (SREBP)/SREBP cleavage-activating protein pathway. *J Biol Chem 275*, 14367-14374.

Lange, Y. (1991). Disposition of intracellular cholesterol in human fibroblasts. *J Lipid Res 32*, 329-339.

Lange, Y., Swaisgood, M.H., Ramos, B.V., and Steck, T.L. (1989). Plasma membranes contain half the phospholipid and 90% of the cholesterol and sphingomyelin in cultured human fibroblasts. *J Biol Chem 264*, 3786-3793.

Langmann, T., Klucken, J., Reil, M., Liebisch, G., Luciani, M.F., Chimini, G., Kaminski, W.E., and Schmitz, G. (1999). Molecular cloning of the human ATP-binding cassette transporter 1 (hABC1): evidence for sterol-dependent regulation in macrophages. *Biochem Biophys Res Commun 257*, 29-33.

Lapierre, L.A., Tuma, P.L., Navarre, J., Goldenring, J.R., and Anderson, J.M. (1999). VAP-33 localizes to both an intracellular vesicle population and with occludin at the tight junction. *J Cell Sci 112 (Pt 21)*, 3723-3732.

Lee, I., Doray, B., Govero, J., and Kornfeld, S. (2008). Binding of cargo sorting signals to AP-1 enhances its association with ADP ribosylation factor 1-GTP. *J Cell Biol 180*, 467-472.

Lee, M.C., Orci, L., Hamamoto, S., Futai, E., Ravazzola, M., and Schekman, R. (2005). Sar1p N-terminal helix initiates membrane curvature and completes the fission of a COPII vesicle. *Cell 122*, 605-617.

Lefrancois, S., and McCormick, P.J. (2007). The Arf GEF GBF1 is required for GGA recruitment to Golgi membranes. *Traffic 8*, 1440-1451.

Lemmon, M.A. (2008). Membrane recognition by phospholipid-binding domains. *Nat Rev Mol Cell Biol 9*, 99-111.

Lenoir, M., Coskun, U., Grzybek, M., Cao, X., Buschhorn, S.B., James, J., Simons, K., and Overduin, M. (2010). Structural basis of wedging the Golgi membrane by FAPP pleckstrin homology domains. *EMBO Rep 11*, 279-284.

Lessmann, E., Ngo, M., Leitges, M., Minguet, S., Ridgway, N.D., and Huber, M. (2007). Oxysterol-binding protein-related protein (ORP) 9 is a PDK-2 substrate and regulates Akt phosphorylation. *Cell Signal 19*, 384-392.

Lev, S. (2004). The role of the Nir/rdgB protein family in membrane trafficking and cytoskeleton remodeling. *Exp Cell Res 297*, 1-10.

Lev, S. (2006). Lipid homoeostasis and Golgi secretory function. *Biochem Soc Trans 34*, 363-366.

Lev, S., Ben Halevy, D., Peretti, D., and Dahan, N. (2008). The VAP protein family: from cellular functions to motor neuron disease. *Trends Cell Biol 18*, 282-290.

Lev, S., Hernandez, J., Martinez, R., Chen, A., Plowman, G., and Schlessinger, J. (1999). Identification of a novel family of targets of PYK2 related to Drosophila retinal degeneration B (rdgB) protein. *Mol Cell Biol 19*, 2278-2288.

Levanon, D., Hsieh, C.L., Francke, U., Dawson, P.A., Ridgway, N.D., Brown, M.S., and Goldstein, J.L. (1990). cDNA cloning of human oxysterol-binding protein and localization of the gene to human chromosome 11 and mouse chromosome 19. *Genomics 7*, 65-74.

Levine, T. (2004). Short-range intracellular trafficking of small molecules across endoplasmic reticulum junctions. *Trends Cell Biol 14*, 483-490.

Li, Z., Hailemariam, T.K., Zhou, H., Li, Y., Duckworth, D.C., Peake, D.A., Zhang, Y., Kuo, M.S., Cao, G., and Jiang, X.C. (2007). Inhibition of sphingomyelin synthase (SMS) affects intracellular sphingomyelin accumulation and plasma membrane lipid organization. *Biochim Biophys Acta 1771*, 1186-1194.

Litvak, V., Dahan, N., Ramachandran, S., Sabanay, H., and Lev, S. (2005). Maintenance of the diacylglycerol level in the Golgi apparatus by the Nir2 protein is critical for Golgi secretory function. *Nat Cell Biol 7*, 225-234.

Litvak, V., Shaul, Y.D., Shulewitz, M., Amarilio, R., Carmon, S., and Lev, S. (2002). Targeting of Nir2 to lipid droplets is regulated by a specific threonine residue within its PI-transfer domain. *Curr Biol 12*, 1513-1518.

Liu, X., Zhang, C., Xing, G., Chen, Q., and He, F. (2001). Functional characterization of novel human ARFGAP3. *FEBS Lett 490*, 79-83.

Liu, Y., Kahn, R.A., and Prestegard, J.H. (2009). Structure and membrane interaction of myristoylated ARF1. *Structure 17*, 79-87.

Loewen, C.J., Roy, A., and Levine, T.P. (2003). A conserved ER targeting motif in three families of lipid binding proteins and in Opi1p binds VAP. *EMBO J 22*, 2025-2035.

Lupashin, V., and Sztul, E. (2005). Golgi tethering factors. *Biochim Biophys Acta 1744*, 325-339.

Malinin, V.S., Frederik, P., and Lentz, B.R. (2002). Osmotic and curvature stress affect PEG-induced fusion of lipid vesicles but not mixing of their lipids. *Biophys J 82*, 2090-2100.

Manneville, J.B., Casella, J.F., Ambroggio, E., Gounon, P., Bertherat, J., Bassereau, P., Cartaud, J., Antonny, B., and Goud, B. (2008). COPI coat assembly occurs on liquid-disordered domains and the associated membrane deformations are limited by membrane tension. *Proc Natl Acad Sci U S A 105*, 16946-16951.

Manolea, F., Claude, A., Chun, J., Rosas, J., and Melancon, P. (2008). Distinct functions for Arf guanine nucleotide exchange factors at the Golgi complex: GBF1 and BIGs are required for assembly and maintenance of the Golgi stack and trans-Golgi network, respectively. *Mol Biol Cell 19*, 523-535.

Marquardt, T., and Denecke, J. (2003). Congenital disorders of glycosylation: review of their molecular bases, clinical presentations and specific therapies. *Eur J Pediatr 162*, 359-379.

Matzaris, M., O'Malley, C.J., Badger, A., Speed, C.J., Bird, P.I., and Mitchell, C.A. (1998). Distinct membrane and cytosolic forms of inositol polyphosphate 5-phosphatase II. Efficient membrane localization requires two discrete domains. *J Biol Chem 273*, 8256-8267.

Mayer, A., Scheglmann, D., Dove, S., Glatz, A., Wickner, W., and Haas, A. (2000). Phosphatidylinositol 4,5-bisphosphate regulates two steps of homotypic vacuole fusion. *Mol Biol Cell 11*, 807-817.

Miller, S.E., Collins, B.M., McCoy, A.J., Robinson, M.S., and Owen, D.J. (2007). A SNARE-adaptor interaction is a new mode of cargo recognition in clathrin-coated vesicles. *Nature 450*, 570-574.

Mills, I.G., Praefcke, G.J., Vallis, Y., Peter, B.J., Olesen, L.E., Gallop, J.L., Butler, P.J., Evans, P.R., and McMahon, H.T. (2003). EpsinR: an AP1/clathrin interacting protein involved in vesicle trafficking. *J Cell Biol 160*, 213-222.

Mima, J., Hickey, C.M., Xu, H., Jun, Y., and Wickner, W. (2008). Reconstituted membrane fusion requires regulatory lipids, SNAREs and synergistic SNARE chaperones. *EMBO J 27*, 2031-2042.

Mironov, A., Colanzi, A., Silletta, M.G., Fiucci, G., Flati, S., Fusella, A., Polishchuk, R., Mironov, A., Jr., Di Tullio, G., Weigert, R., Malhotra, V., Corda, D., De Matteis, M.A., and Luini, A. (1997). Role of NAD+ and ADP-ribosylation in the maintenance of the Golgi structure. *J Cell Biol 139*, 1109-1118.

Mitchell, C.A., Connolly, T.M., and Majerus, P.W. (1989). Identification and isolation of a 75-kDa inositol polyphosphate-5-phosphatase from human platelets. *J Biol Chem 264*, 8873-8877.

Mohammadi, A., Perry, R.J., Storey, M.K., Cook, H.W., Byers, D.M., and Ridgway, N.D. (2001). Golgi localization and phosphorylation of oxysterol binding protein in Niemann-Pick C and U18666A-treated cells. *J Lipid Res 42*, 1062-1071.

Moreau, P., Cassagne, C., Keenan, T.W., and Morre, D.J. (1993). Ceramide excluded from cell-free vesicular lipid transfer from endoplasmic reticulum to Golgi apparatus. Evidence for lipid sorting. *Biochim Biophys Acta 1146*, 9-16.

Morikawa, R.K., Aoki, J., Kano, F., Murata, M., Yamamoto, A., Tsujimoto, M., and Arai, H. (2009). Intracellular phospholipase A1gamma (iPLA1gamma) is a novel factor involved in coat protein complex I- and Rab6-independent retrograde transport between the endoplasmic reticulum and the Golgi complex. *J Biol Chem 284*, 26620-26630.

Mukherjee, S., and Maxfield, F.R. (2004). Lipid and cholesterol trafficking in NPC. *Biochim Biophys Acta 1685*, 28-37.

Nagao, K., Takahashi, K., Hanada, K., Kioka, N., Matsuo, M., and Ueda, K. (2007). Enhanced apoA-I-dependent cholesterol efflux by ABCA1 from sphingomyelin-deficient Chinese hamster ovary cells. *J Biol Chem 282*, 14868-14874.

Nardini, M., Spano, S., Cericola, C., Pesce, A., Massaro, A., Millo, E., Luini, A., Corda, D., and Bolognesi, M. (2003). CtBP/BARS: a dual-function protein involved in transcription co-repression and Golgi membrane fission. *EMBO J 22*, 3122-3130.

Natarajan, P., Liu, K., Patil, D.V., Sciorra, V.A., Jackson, C.L., and Graham, T.R. (2009). Regulation of a Golgi flippase by phosphoinositides and an ArfGEF. *Nat Cell Biol 11*, 1421-1426.

Nemoto, Y., Kearns, B.G., Wenk, M.R., Chen, H., Mori, K., Alb, J.G., Jr., De Camilli, P., and Bankaitis, V.A. (2000). Functional characterization of a mammalian Sac1 and mutants exhibiting substrate-specific defects in phosphoinositide phosphatase activity. *J Biol Chem 275*, 34293-34305.

Ngo, M., and Ridgway, N.D. (2009). Oxysterol binding protein-related Protein 9 (ORP9) is a cholesterol transfer protein that regulates Golgi structure and function. *Mol Biol Cell 20*, 1388-1399.

Ngo, M.H., Colbourne, T.R., and Ridgway, N.D. (2010). Functional implications of sterol transport by the oxysterol-binding protein gene family. *Biochem J 429*, 13-24.

Nhek, S., Ngo, M., Yang, X., Ng, M.M., Field, S.J., Asara, J.M., Ridgway, N.D., and Toker, A. (2010). Regulation of OSBP Golgi Localization through Protein Kinase D-mediated Phosphorylation. *Mol Biol Cell*.

Nogi, T., Shiba, Y., Kawasaki, M., Shiba, T., Matsugaki, N., Igarashi, N., Suzuki, M., Kato, R., Takatsu, H., Nakayama, K., and Wakatsuki, S. (2002). Structural basis for the accessory protein recruitment by the gamma-adaptin ear domain. *Nat Struct Biol 9*, 527-531.

Ohgami, N., Ko, D.C., Thomas, M., Scott, M.P., Chang, C.C., and Chang, T.Y. (2004). *Binding between the Niemann-Pick C1 protein and a photoactivatable cholesterol analog requires a functional sterol-sensing domain.* Proc Natl Acad Sci U S A *101*, 12473-12478.

Olivos-Glander, I.M., Janne, P.A., and Nussbaum, R.L. (1995). The oculocerebrorenal syndrome gene product is a 105-kD protein localized to the Golgi complex. *Am J Hum Genet 57*, 817-823.

Orci, L., Montesano, R., Meda, P., Malaisse-Lagae, F., Brown, D., Perrelet, A., and Vassalli, P. (1981). *Heterogeneous distribution of filipin--cholesterol complexes across the cisternae of the Golgi apparatus*. Proc Natl Acad Sci U S A *78*, 293-297.

Orso, E., Broccardo, C., Kaminski, W.E., Bottcher, A., Liebisch, G., Drobnik, W., Gotz, A., Chambenoit, O., Diederich, W., Langmann, T., Spruss, T., Luciani, M.F., Rothe, G., Lackner, K.J., Chimini, G., and Schmitz, G. (2000). Transport of lipids from golgi to plasma membrane is defective in tangier disease patients and Abc1-deficient mice. *Nat Genet 24*, 192-196.

Pathre, P., Shome, K., Blumental-Perry, A., Bielli, A., Haney, C.J., Alber, S., Watkins, S.C., Romero, G., and Aridor, M. (2003). Activation of phospholipase D by the small GTPase Sar1p is required to support COPII assembly and ER export. *EMBO J 22*, 4059-4069.

Peden, A.A., Oorschot, V., Hesser, B.A., Austin, C.D., Scheller, R.H., and Klumperman, J. (2004). Localization of the AP-3 adaptor complex defines a novel endosomal exit site for lysosomal membrane proteins. *J Cell Biol 164*, 1065-1076.

Peretti, D., Dahan, N., Shimoni, E., Hirschberg, K., and Lev, S. (2008). Coordinated Lipid Transfer between the Endoplasmic Reticulum and the Golgi Complex Requires the VAP Proteins and Is Essential for Golgi-mediated Transport. *Mol Biol Cell.19*, 3871-3884.

Perry, R.J., and Ridgway, N.D. (2006). Oxysterol-binding protein and vesicle-associated membrane protein-associated protein are required for sterol-dependent activation of the ceramide transport protein. *Mol Biol Cell 17*, 2604-2616.

Pinton, P., Pozzan, T., and Rizzuto, R. (1998). The Golgi apparatus is an inositol 1,4,5-trisphosphate-sensitive Ca^{2+} store, with functional properties distinct from those of the endoplasmic reticulum. *EMBO J 17*, 5298-5308.

Ponting, C.P., and Aravind, L. (1999). START: a lipid-binding domain in StAR, HD-ZIP and signalling proteins. *Trends Biochem Sci 24*, 130-132.

Porn, M.I., and Slotte, J.P. (1990). Reversible effects of sphingomyelin degradation on cholesterol distribution and metabolism in fibroblasts and transformed neuroblastoma cells. *Biochem J 271*, 121-126.

Puthenveedu, M.A., and Linstedt, A.D. (2005). Subcompartmentalizing the Golgi apparatus. *Curr Opin Cell Biol 17*, 369-375.

Radhakrishnan, A., Goldstein, J.L., McDonald, J.G., and Brown, M.S. (2008). Switch-like control of SREBP-2 transport triggered by small changes in ER cholesterol: a delicate balance. *Cell Metab 8*, 512-521.

Radhakrishnan, A., Ikeda, Y., Kwon, H.J., Brown, M.S., and Goldstein, J.L. (2007). *Sterol-regulated transport of SREBPs from endoplasmic reticulum to Golgi: oxysterols block transport by binding to Insig*. Proc Natl Acad Sci U S A *104*, 6511-6518.

Rao, R.P., Yuan, C., Allegood, J.C., Rawat, S.S., Edwards, M.B., Wang, X., Merrill, A.H., Jr., Acharya, U., and Acharya, J.K. (2007). *Ceramide transfer protein function is essential for normal oxidative stress response and lifespan.* Proc Natl Acad Sci U S A *104*, 11364-11369.

Raya, A., Revert-Ros, F., Martinez-Martinez, P., Navarro, S., Rosello, E., Vieites, B., Granero, F., Forteza, J., and Saus, J. (2000). Goodpasture antigen-binding protein, the kinase that phosphorylates the goodpasture antigen, is an alternatively spliced variant implicated in autoimmune pathogenesis. *J Biol Chem 275*, 40392-40399.

Reddy, J.V., Burguete, A.S., Sridevi, K., Ganley, I.G., Nottingham, R.M., and Pfeffer, S.R. (2006). A functional role for the GCC185 golgin in mannose 6-phosphate receptor recycling. *Mol Biol Cell 17*, 4353-4363.

Resh, M.D. (1999). Fatty acylation of proteins: new insights into membrane targeting of myristoylated and palmitoylated proteins. *Biochim Biophys Acta 1451*, 1-16.

Ridgway, N.D. (2000). Interactions between metabolism and intracellular distribution of cholesterol and sphingomyelin. *Biochim Biophys Acta 1484*, 129-141.

Ridgway, N.D., Dawson, P.A., Ho, Y.K., Brown, M.S., and Goldstein, J.L. (1992). Translocation of oxysterol binding protein to Golgi apparatus triggered by ligand binding. *J Cell Biol 116*, 307-319.

Ridgway, N.D., and Lagace, T.A. (2003). Regulation of the CDP-choline pathway by sterol regulatory element binding proteins involves transcriptional and post-transcriptional mechanisms. *Biochem J 372*, 811-819.

Ridsdale, A., Denis, M., Gougeon, P.Y., Ngsee, J.K., Presley, J.F., and Zha, X. (2006). Cholesterol is required for efficient endoplasmic reticulum-to-Golgi transport of secretory membrane proteins. *Mol Biol Cell 17*, 1593-1605.

Riebeling, C., Morris, A.J., and Shields, D. (2009). Phospholipase D in the Golgi apparatus. *Biochim Biophys Acta 1791*, 876-880.

Rietveld, A., Neutz, S., Simons, K., and Eaton, S. (1999). Association of sterol- and glycosylphosphatidylinositol-linked proteins with Drosophila raft lipid microdomains. *J Biol Chem 274*, 12049-12054.

Robinson, M.S. (2004). Adaptable adaptors for coated vesicles. Trends Cell Biol *14*, 167-174.

Roper, K., Corbeil, D., and Huttner, W.B. (2000). Retention of prominin in microvilli reveals distinct cholesterol-based lipid micro-domains in the apical plasma membrane. *Nat Cell Biol 2*, 582-592.

Runz, H., Miura, K., Weiss, M., and Pepperkok, R. (2006). Sterols regulate ER-export dynamics of secretory cargo protein ts-O45-G. *EMBO J 25*, 2953-2965.

Rusten, T.E., and Stenmark, H. (2006). Analyzing phosphoinositides and their interacting proteins. *Nat Methods 3*, 251-258.

Saito, S., Matsui, H., Kawano, M., Kumagai, K., Tomishige, N., Hanada, K., Echigo, S., Tamura, S., and Kobayashi, T. (2008). Protein phosphatase 2Cepsilon is an endoplasmic reticulum integral membrane protein that dephosphorylates the ceramide transport protein CERT to enhance its association with organelle membranes. *J Biol Chem 283*, 6584-6593.

San Pietro, E., Capestrano, M., Polishchuk, E.V., DiPentima, A., Trucco, A., Zizza, P., Mariggio, S., Pulvirenti, T., Sallese, M., Tete, S., Mironov, A.A., Leslie, C.C., Corda, D., Luini, A., and Polishchuk, R.S. (2009). Group IV phospholipase A(2)alpha controls the formation of inter-cisternal continuities involved in intra-Golgi transport. *PLoS Biol 7*, e1000194.

Scheek, S., Brown, M.S., and Goldstein, J.L. (1997). *Sphingomyelin depletion in cultured cells blocks proteolysis of sterol regulatory element binding proteins at site 1*. Proc Natl Acad Sci U S A *94*, 11179-11183.

Schmidt, J.A., and Brown, W.J. (2009). Lysophosphatidic acid acyltransferase 3 regulates Golgi complex structure and function. *J Cell Biol 186*, 211-218.

Scott, C., and Ioannou, Y.A. (2004). The NPC1 protein: structure implies function. *Biochim Biophys Acta 1685*, 8-13.

Sever, N., Yang, T., Brown, M.S., Goldstein, J.L., and DeBose-Boyd, R.A. (2003). Accelerated degradation of HMG CoA reductase mediated by binding of insig-1 to its sterol-sensing domain. *Mol Cell 11*, 25-33.

Shindou, H., and Shimizu, T. (2009). Acyl-CoA:lysophospholipid acyltransferases. *J Biol Chem 284*, 1-5.

Shinotsuka, C., Waguri, S., Wakasugi, M., Uchiyama, Y., and Nakayama, K. (2002). Dominant-negative mutant of BIG2, an ARF-guanine nucleotide exchange factor, specifically affects membrane trafficking from the trans-Golgi network through inhibiting membrane association of AP-1 and GGA coat proteins. *Biochem Biophys Res Commun 294*, 254-260.

Simons, K., and Toomre, D. (2000). *Lipid rafts and signal transduction*. Nat Rev Mol Cell Biol *1*, 31-39.

Skehel, P.A., Fabian-Fine, R., and Kandel, E.R. (2000). *Mouse VAP33 is associated with the endoplasmic reticulum and microtubules.* Proc Natl Acad Sci U S A *97*, 1101-1106.

Smith, J.D., Waelde, C., Horwitz, A., and Zheng, P. (2002). Evaluation of the role of phosphatidylserine translocase activity in ABCA1-mediated lipid efflux. *J Biol Chem 277*, 17797-17803.

Soccio, R.E., and Breslow, J.L. (2003). StAR-related lipid transfer (START) proteins: mediators of intracellular lipid metabolism. *J Biol Chem 278*, 22183-22186.

Song, B.L., and DeBose-Boyd, R.A. (2004). Ubiquitination of 3-hydroxy-3-methylglutaryl-CoA reductase in permeabilized cells mediated by cytosolic E1 and a putative membrane-bound ubiquitin ligase. *J Biol Chem 279*, 28798-28806.

Sprong, H., Kruithof, B., Leijendekker, R., Slot, J.W., van Meer, G., and van der Sluijs, P. (1998). UDP-galactose:ceramide galactosyltransferase is a class I integral membrane protein of the endoplasmic reticulum. *J Biol Chem 273*, 25880-25888.

Stahelin, R.V., Karathanassis, D., Murray, D., Williams, R.L., and Cho, W. (2007). Structural and membrane binding analysis of the Phox homology domain of Bem1p: basis of phosphatidylinositol 4-phosphate specificity. *J Biol Chem 282*, 25737-25747.

Stenmark, H. (2009). Rab GTPases as coordinators of vesicle traffic. *Nat Rev Mol Cell Biol 10*, 513-525.

Storey, M.K., Byers, D.M., Cook, H.W., and Ridgway, N.D. (1998). Cholesterol regulates oxysterol binding protein (OSBP) phosphorylation and Golgi localization in Chinese hamster ovary cells: correlation with stimulation of sphingomyelin synthesis by 25-hydroxycholesterol. *Biochem J 336 (Pt 1)*, 247-256.

Sturley, S.L., Patterson, M.C., Balch, W., and Liscum, L. (2004). The pathophysiology and mechanisms of NP-C disease. *Biochim Biophys Acta 1685*, 83-87.

Suchanek, M., Hynynen, R., Wohlfahrt, G., Lehto, M., Johansson, M., Saarinen, H., Radzikowska, A., Thiele, C., and Olkkonen, V.M. (2007). The mammalian oxysterol-binding protein-related proteins (ORPs) bind 25-hydroxycholesterol in an evolutionarily conserved pocket. *Biochem J 405*, 473-480.

Suchy, S.F., and Nussbaum, R.L. (2002). The deficiency of PIP2 5-phosphatase in Lowe syndrome affects actin polymerization. *Am J Hum Genet 71*, 1420-1427.

Sung, T.C., Altshuller, Y.M., Morris, A.J., and Frohman, M.A. (1999). Molecular analysis of mammalian phospholipase D2. *J Biol Chem 274*, 494-502.

Sung, T.C., Roper, R.L., Zhang, Y., Rudge, S.A., Temel, R., Hammond, S.M., Morris, A.J., Moss, B., Engebrecht, J., and Frohman, M.A. (1997). Mutagenesis of phospholipase D defines a superfamily including a trans-Golgi viral protein required for poxvirus pathogenicity. *EMBO J 16*, 4519-4530.

Sweeney, D.A., Siddhanta, A., and Shields, D. (2002). Fragmentation and re-assembly of the Golgi apparatus in vitro. A requirement for phosphatidic acid and phosphatidylinositol 4,5-bisphosphate synthesis. *J Biol Chem 277*, 3030-3039.

Tafesse, F.G., Huitema, K., Hermansson, M., van der Poel, S., van den Dikkenberg, J., Uphoff, A., Somerharju, P., and Holthuis, J.C. (2007). Both sphingomyelin synthases SMS1 and SMS2 are required for sphingomyelin homeostasis and growth in human HeLa cells. *J Biol Chem 282*, 17537-17547.

Takahashi, K., Kimura, Y., Kioka, N., Matsuo, M., and Ueda, K. (2006). Purification and ATPase activity of human ABCA1. *J Biol Chem 281*, 10760-10768.

Ternes, P., Brouwers, J.F., van den Dikkenberg, J., and Holthuis, J.C. (2009). Sphingomyelin synthase SMS2 displays dual activity as ceramide phosphoethanolamine synthase. *J Lipid Res 50*, 2270-2277.

Toth, B., Balla, A., Ma, H., Knight, Z.A., Shokat, K.M., and Balla, T. (2006). Phosphatidylinositol 4-kinase IIIbeta regulates the transport of ceramide between the endoplasmic reticulum and Golgi. *J Biol Chem 281*, 36369-36377.

Traub, L.M., Ostrom, J.A., and Kornfeld, S. (1993). Biochemical dissection of AP-1 recruitment onto Golgi membranes. *J Cell Biol 123*, 561-573.

Ungewickell, A., Ward, M.E., Ungewickell, E., and Majerus, P.W. (2004). *The inositol polyphosphate 5-phosphatase Ocrl associates with endosomes that are partially coated with clathrin*. Proc Natl Acad Sci U S A *101*, 13501-13506.

Ungewickell, A.J., and Majerus, P.W. (1999). *Increased levels of plasma lysosomal enzymes in patients with Lowe syndrome*. Proc Natl Acad Sci U S A *96*, 13342-13344.

Urano, Y., Watanabe, H., Murphy, S.R., Shibuya, Y., Geng, Y., Peden, A.A., Chang, C.C., and Chang, T.Y. (2008). *Transport of LDL-derived cholesterol from the NPC1 compartment to the ER involves the trans-Golgi network and the SNARE protein complex.* Proc Natl Acad Sci U S A *105*, 16513-16518.

van Meer, G., Voelker, D.R., and Feigenson, G.W. (2008). Membrane lipids: where they are and how they behave. *Nat Rev Mol Cell Biol 9*, 112-124.

Vance, J.E., and Vance, D.E. (2004). Phospholipid biosynthesis in mammalian cells. *Biochem Cell Biol 82*, 113-128.

Vanhaesebroeck, B., Leevers, S.J., Ahmadi, K., Timms, J., Katso, R., Driscoll, P.C., Woscholski, R., Parker, P.J., and Waterfield, M.D. (2001). Synthesis and function of 3-phosphorylated inositol lipids. *Annu Rev Biochem 70*, 535-602.

Villani, M., Subathra, M., Im, Y.B., Choi, Y., Signorelli, P., Del Poeta, M., and Luberto, C. (2008). Sphingomyelin synthases regulate production of diacylglycerol at the Golgi. *Biochem J 414*, 31-41.

Vordtriede, P.B., Doan, C.N., Tremblay, J.M., Helmkamp, G.M., Jr., and Yoder, M.D. (2005). Structure of PITPβeta in complex with phosphatidylcholine: comparison of structure and lipid transfer to other PITP isoforms. *Biochemistry 44*, 14760-14771.

Wahle, T., Prager, K., Raffler, N., Haass, C., Famulok, M., and Walter, J. (2005). GGA proteins regulate retrograde transport of BACE1 from endosomes to the trans-Golgi network. *Mol Cell Neurosci 29*, 453-461.

Wang, E., Norred, W.P., Bacon, C.W., Riley, R.T., and Merrill, A.H., Jr. (1991). Inhibition of sphingolipid biosynthesis by fumonisins. Implications for diseases associated with Fusarium moniliforme. *J Biol Chem 266*, 14486-14490.

Wang, J., Sun, H.Q., Macia, E., Kirchhausen, T., Watson, H., Bonifacino, J.S., and Yin, H.L. (2007). PI4P promotes the recruitment of the GGA adaptor proteins to the trans-Golgi network and regulates their recognition of the ubiquitin sorting signal. *Mol Biol Cell 18*, 2646-2655.

Wang, M.L., Motamed, M., Infante, R.E., Abi-Mosleh, L., Kwon, H.J., Brown, M.S., and Goldstein, J.L. (2010). Identification of surface residues on niemann-pick C2 essential for hydrophobic Handoff of cholesterol to NPC1 in lysosomes. *Cell Metab 12*, 166-173.

Wang, P.Y., Weng, J., and Anderson, R.G. (2005). OSBP is a cholesterol-regulated scaffolding protein in control of ERK 1/2 activation. *Science 307*, 1472-1476.

Wang, P.Y., Weng, J., Lee, S., and Anderson, R.G. (2008). The N terminus controls sterol binding while the C terminus regulates the scaffolding function of OSBP. *J Biol Chem 283*, 8034-8045.

Wang, W., Sacher, M., and Ferro-Novick, S. (2000a). TRAPP stimulates guanine nucleotide exchange on Ypt1p. *J Cell Biol 151*, 289-296.

Wang, Y., Thiele, C., and Huttner, W.B. (2000b). Cholesterol is required for the formation of regulated and constitutive secretory vesicles from the trans-Golgi network. *Traffic 1*, 952-962.

Wang, Y.J., Wang, J., Sun, H.Q., Martinez, M., Sun, Y.X., Macia, E., Kirchhausen, T., Albanesi, J.P., Roth, M.G., and Yin, H.L. (2003). Phosphatidylinositol 4 phosphate regulates targeting of clathrin adaptor AP-1 complexes to the Golgi. *Cell 114*, 299-310.

Warnock, D.E., Roberts, C., Lutz, M.S., Blackburn, W.A., Young, W.W., Jr., and Baenziger, J.U. (1993). Determination of plasma membrane lipid mass and composition in cultured Chinese hamster ovary cells using high gradient magnetic affinity chromatography. *J Biol Chem 268*, 10145-10153.

Watson, P.J., Frigerio, G., Collins, B.M., Duden, R., and Owen, D.J. (2004). Gamma-COP appendage domain - structure and function. *Traffic 5*, 79-88.

Watt, S.A., Kular, G., Fleming, I.N., Downes, C.P., and Lucocq, J.M. (2002). Subcellular localization of phosphatidylinositol 4,5-bisphosphate using the pleckstrin homology domain of phospholipase C delta1. *Biochem J 363*, 657-666.

Waugh, M.G., Minogue, S., Chotai, D., Berditchevski, F., and Hsuan, J.J. (2006). Lipid and peptide control of phosphatidylinositol 4-kinase IIalpha activity on Golgi-endosomal Rafts. *J Biol Chem 281*, 3757-3763.

Weixel, K.M., Blumental-Perry, A., Watkins, S.C., Aridor, M., and Weisz, O.A. (2005). Distinct Golgi populations of phosphatidylinositol 4-phosphate regulated by phosphatidylinositol 4-kinases. *J Biol Chem 280*, 10501-10508.

Wendel, A.A., Lewin, T.M., and Coleman, R.A. (2009). Glycerol-3-phosphate acyltransferases: rate limiting enzymes of triacylglycerol biosynthesis. *Biochim Biophys Acta 1791*, 501-506.

Wennekes, T., van den Berg, R.J., Boot, R.G., van der Marel, G.A., Overkleeft, H.S., and Aerts, J.M. (2009). Glycosphingolipids--nature, function, and pharmacological modulation. *Angew Chem Int Ed Engl 48*, 8848-8869.

Whisstock, J.C., Wiradjaja, F., Waters, J.E., and Gurung, R. (2002). The structure and function of catalytic domains within inositol polyphosphate 5-phosphatases. *IUBMB Life* 53, 15-23.

Williams, C., Choudhury, R., McKenzie, E., and Lowe, M. (2007). Targeting of the type II inositol polyphosphate 5-phosphatase INPP5B to the early secretory pathway. *J Cell Sci* 120, 3941-3951.

Worgall, T.S., Juliano, R.A., Seo, T., and Deckelbaum, R.J. (2004). Ceramide synthesis correlates with the posttranscriptional regulation of the sterol-regulatory element-binding protein. *Arterioscler Thromb Vasc Biol 24*, 943-948.

Wu, X., Steet, R.A., Bohorov, O., Bakker, J., Newell, J., Krieger, M., Spaapen, L., Kornfeld, S., and Freeze, H.H. (2004). Mutation of the COG complex subunit gene COG7 causes a lethal congenital disorder. *Nat Med 10*, 518-523.

Wyles, J.P., McMaster, C.R., and Ridgway, N.D. (2002). Vesicle-associated membrane protein-associated protein-A (VAP-A) interacts with the oxysterol-binding protein to modify export from the endoplasmic reticulum. *J Biol Chem 277*, 29908-29918.

Xu, Z., Farver, W., Kodukula, S., and Storch, J. (2008). Regulation of sterol transport between membranes and NPC2. *Biochemistry 47*, 11134-11143.

Yamaji, T., Kumagai, K., Tomishige, N., and Hanada, K. (2008). Two sphingolipid transfer proteins, CERT and FAPP2: their roles in sphingolipid metabolism. *IUBMB Life 60*, 511-518.

Yamasaki, A., Menon, S., Yu, S., Barrowman, J., Meerloo, T., Oorschot, V., Klumperman, J., Satoh, A., and Ferro-Novick, S. (2009). mTrs130 is a component of a mammalian TRAPPII complex, a Rab1 GEF that binds to COPI-coated vesicles. *Mol Biol Cell 20*, 4205-4215.

Yang, J.S., Gad, H., Lee, S.Y., Mironov, A., Zhang, L., Beznoussenko, G.V., Valente, C., Turacchio, G., Bonsra, A.N., Du, G., Baldanzi, G., Graziani, A., Bourgoin, S., Frohman, M.A., Luini, A., and Hsu, V.W. (2008). A role for phosphatidic acid in COPI vesicle fission yields insights into Golgi maintenance. *Nat Cell Biol 10*, 1146-1153.

Yang, J.S., Lee, S.Y., Spano, S., Gad, H., Zhang, L., Nie, Z., Bonazzi, M., Corda, D., Luini, A., and Hsu, V.W. (2005). A role for BARS at the fission step of COPI vesicle formation from Golgi membrane. *EMBO J 24*, 4133-4143.

Yeaman, C., Ayala, M.I., Wright, J.R., Bard, F., Bossard, C., Ang, A., Maeda, Y., Seufferlein, T., Mellman, I., Nelson, W.J., and Malhotra, V. (2004). Protein kinase D regulates basolateral membrane protein exit from trans-Golgi network. *Nat Cell Biol 6*, 106-112.

Ying, M., Grimmer, S., Iversen, T.G., Van Deurs, B., and Sandvig, K. (2003). Cholesterol loading induces a block in the exit of VSVG from the TGN. *Traffic 4*, 772-784.

Zerbinatti, C.V., Cordy, J.M., Chen, C.D., Guillily, M., Suon, S., Ray, W.J., Seabrook, G.R., Abraham, C.R., and Wolozin, B. (2008). Oxysterol-binding protein-1 (OSBP1) modulates processing and trafficking of the amyloid precursor protein. *Mol Neurodegener 3*, 5.

Zhao, C., Slevin, J.T., and Whiteheart, S.W. (2007). Cellular functions of NSF: not just SNAPs and SNAREs. *FEBS Lett 581*, 2140-2149.

Zhong, P., Chen, Y.A., Tam, D., Chung, D., Scheller, R.H., and Miljanich, G.P. (1997). An alpha-helical minimal binding domain within the H3 domain of syntaxin is required for SNAP-25 binding. *Biochemistry 36*, 4317-4326.

Zhou, Y., Li, S., Mayranpaa, M.I., Zhong, W., Back, N., Yan, D., and Olkkonen, V.M. (2010). OSBP-related protein 11 (ORP11) dimerizes with ORP9 and localizes at the Golgi-late endosome interface. *Exp Cell Res. 316,* 3304-3316.

Zimmerberg, J., and Kozlov, M.M. (2006). How proteins produce cellular membrane curvature. *Nat Rev Mol Cell Biol 7*, 9-19.

Chapter 3

SIGNALING PATHWAYS CONTROLLING MITOTIC GOLGI BREAKDOWN IN MAMMALIAN CELLS

Inmaculada López-Sánchez and Pedro A. Lazo[*]

Experimental Therapeutics and Translational Oncology Program Instituto de Biología Molecular y Celular del Cáncer, Consejo Superior de Investigaciones Científicas (CSIC), Universidad de Salamanca, Campus Miguel de Unamuno, E-37007 Salamanca, Spain.

ABSTRACT

In mitosis, each daughter cell must receive a complete and equal set of cellular components. Cellular organelles that are single copy, such as endoplasmic reticulum, nuclear envelope and Golgi apparatus, have to break down to allow their correct distribution between daughter cells. The mammalian Golgi is a continuous membranous system formed by cistern stacks, tubules and small vesicles that are located in the perinuclear area. At the onset of mitosis, the Golgi apparatus undergoes a sequential fragmentation that is highly coordinated with mitotic progression and in which reversible phosphorylation plays a critical regulatory role. In fact, several kinases have been implicated in each stage of this fragmentation process. Before mitotic disassembly, the lateral connections between the stacks are severed resulting in the formation of isolated cisternae. Several kinases such as mitogen-activated protein kinase kinase 1(MEK1), Raf-1, ERK1c, ERK2, Plk3, VRK1, several Golgi matrix proteins (GRASP65 and GRASP55) and the membrane fission protein BARS have been shown to mediate signals in this first step that takes place in late G2 phase. As prophase progresses, the isolated cisternae are first unstacked followed by its breakage into smaller vesicles and tubules that accumulate around the two spindle poles at metaphase. Unstacking and vesiculation are triggered by several proteins including kinases (Plk1 and Cdc2), the GTPase ARF-1 and inactivation of membrane fusion complexes (VCP and NSF). Post-mitotic Golgi reassembly consists of two processes: membrane fusion mediated by two ATPases, VCP and NSF; and cistern restacking mediated by dephosphorylation of Golgi matrix proteins (GRASP65 and GM130) by phosphatase PP2A (Bα). Apart from the tight regulation by reversible phosphorylation, it seems that mitotic Golgi membrane dynamics also involves a cycle of

---

[*] Centro de Investigación del Cáncer, CSIC- Universidad de Salamanca, Campus Miguel de Unamuno, E-37007 Salamanca, Spain. Dr. P. A. Lazo, Tel: 34 923 294 804, Fax: 34 923 294 795, E-mail: plazozbi@usal.es

ubiquitination during disassembly and deubiquitination during reassembly in part regulated by the VCP-mediated pathway.

1. INTRODUCTION

The functions of the Golgi apparatus are conserved throughout eukaryotic evolution but its morphology and spatial organization vary among organisms. In *Saccharomyces cerevisae* individual Golgi cisternae are scattered throughout the cytoplasm [1]; whereas in most eukaryotic organisms, such as in plants, cisternae are arranged into ordered stacks and the Golgi is made of many individual dispersed stacks [2]. In mammalian cells, Golgi stacks are laterally linked forming the characteristic ribbon-like complex that is located in the perinuclear region next to the centrosome [3]. In mammalian cells, cellular organelles that exist as a single copy such as the endoplasmic reticulum, the nuclear envelope and the Golgi apparatus have to break down in mitosis to allow its correct partitioning between daughter cells. In the case of the Golgi complex, its continuous membranous system is first laterally severed at late G2 phase, generating isolated Golgi stacks [4, 5]. Then, cisternae unstack and break down into many smaller vesicles and tubules, and disperse around the spindle poles at metaphase forming the "Golgi mitotic haze" [6-9] (Figure 1). After completion of Golgi mitotic fragmentation and partitioning into daughter cells, Golgi vesicles reassemble in a two step process: membrane regrowth and cisternal restacking. The Golgi disassembly and reassembly is tightly coordinated with cell cycle progression, and is regulated by two types of protein modifications, reversible phosphorylation and monoubiquitination. The proteins already associated with the control of Golgi in mitosis include Ser-Thr kinases, phosphatases, and membrane proteins (Table 1). Among them are for example, Raf-1 [10], MEK1 [5, 11-13], ERK1c [14], Cdc2 [15-17], Plk1 [16, 18, 19], ERK2 [5, 20], Plk3 [21], VRK1 [22] and PP2A [23, 24]; Golgi matrix proteins like GM130 [23], GRASP55 [20, 25] and GRASP65 [16, 26, 27], the fission protein CtBP3/BARS [4], the GTPase ARF-1 [28, 29] and proteins related with fusion events such as VCP [30] (Table 1). But all these proteins and their role in Golgi dynamics have to be coordinated with regulatory signals implicated in the control of cell division.

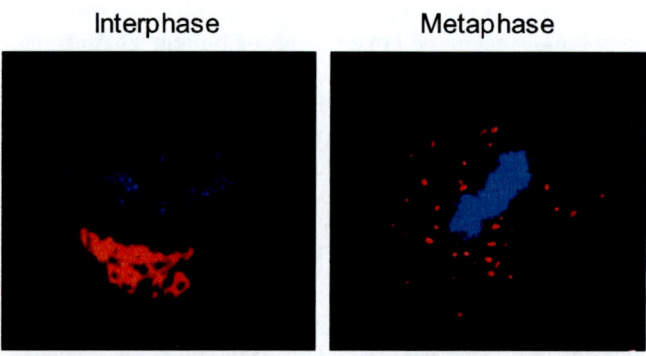

Figure 1. Golgi membranes morphology in interphase and metaphase in HeLa cells. In interphase, the Golgi complex is a continuous membranous system. In metaphase, the Golgi complex is dispersed into vesicles and tubules throughout the cytoplasm around the DNA (mitotic Golgi Haze). The Golgi ribbon is labeled with an anti- Giantin antibody in red and DNA is labeled with DAPI in blue.

Table 1. Proteins involved in the Golgi fragmentation process in mitosis

Protein	Reference
1. Unlinking of the Golgi ribbon in late G2 phase	
Kinases	
Raf-1	(10)
MEK1	(5, 11-13, 21)
ERK2	(5, 20)
ERK1c	(14)
Plk3	(21)
VRK1	(22)
Golgi matrix proteins	
GRASP55	(5, 20, 46, 52)
GRASP65	(19, 44, 46, 80)
Membrane fission protein	
BARS	(4, 48)
2. Golgi stacks disassembly and vesiculation	
Kinases	
Plk1	(16, 19)
Cdc2	(15-17, 19, 44)
GTPases	
ARF-1	(24, 28, 29)
Membrane fusion complexes (ATPases)	
p47 (VCP)	(17)
GM130 (NSF)	(15)

2. UNLINKING OF THE GOLGI RIBBON IN LATE G2 PHASE

Apart from the fact that the Golgi complex disassembly is necessary for its correct partitioning between daughter cells [6, 7, 31], recently, several groups have suggested that the first modification that takes place in late G2 phase and that results in the isolation of Golgi stacks is a prerequisite for mitotic entry [5].

Moreover, that this unlinking of the Golgi ribbon has been proposed as a novel non-DNA checkpoint that is required for G2/M transition, and requires a connection with signaling pathways controlling cell division [5, 9, 32].

MAP kinases canonical functions are regulation of proliferation, differentiation and apoptosis [33].

However, it has been demonstrated that MAP kinases are not only activated in mitogenic cascades related with cell cycle entry, but also during mitosis to regulate several processes. The MEK/ERK signaling pathway, in addition to its classical role, is specifically activated in G2/M transition independently of extracellular growth factor stimulation [10, 34, 35].

Raf-1, for example, seems to be activated by a different mechanism that does not include classical proteins such as Ras, 14-3-3 or Src [36]; in this case MEK1 activation requires other components apart from Raf-1 to be activated in mitosis, since different MEK1 phosphorylations are detected when comparing phosphopeptides from mitotically activated MEK1 with those from Raf-1 activated MEK1 [10, 12].

These differences might represent a subpopulation with different signaling characteristics.

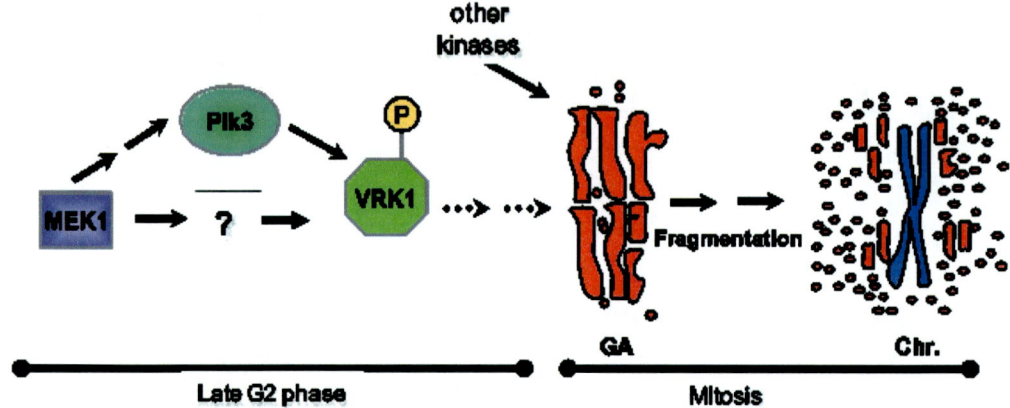

Figure 2. MEK1-Plk3-VRK1 signaling pathway in Golgi fragmentation. Reprinted with permission from American Society for Microbiology (22).

Several kinases downstream of MEK1 have been implicated in mitotic Golgi fragmentation, such as ERK proteins (ERK2 and ERK1c) [14, 20, 37], Plk3 [21] and VRK1 [22] although their exact temporal and spatial role has not yet been completely established. Currently, the role of MEK1 in Golgi ribbon unlinking, occurring in G2, is well characterized. Depletion by RNA interference or inhibition of MEK1 delays, but does not block, mitotic entry, suggesting that MEK1 signaling is required for timely G2/M transition. In accordance with this idea, in MEK1-knockdown cells, treatment with brefeldin A (BFA), a non-competitive inhibitor of ARF-1 that induces dispersal of the Golgi membranes [38, 39], and depletion of GRASP65 can bypass the MEK1 requirement [5]. Therefore, the MEK1-dependent delay is abrogated in cells with Golgi complex already dispersed either by BFA or siGRASP65 [5]. Plk3 is downstream of MEK/ERK cascade since ectopic expression of activated MEK1 results in activation of Plk3. Moreover, activated MEK1 stimulates Golgi breakdown in the presence of Plk3 but not the kinase-defective Plk3^{K52R} [21]. Plk3 transmits part of the signal from MEK1 through VRK1 to induce Golgi breakdown [22] (Figure 2). In agreement with this notion, Plk3 induces Golgi breakdown in the presence of VRK1 but not the kinase-dead VRK1^{K179E} or the active kinase VRK1^{S342A} that is not phosphorylated by Plk3 [22]. These data confirm that VRK1 is downstream of Plk3 and that VRK1 serine 342 is the residue phosphorylated in this signaling pathway [22]. Besides, knocking-down endogenous VRK1 protein also blocks Golgi fragmentation induced either by MEK1 or Plk3, which means that VRK1 is a new step in the already known MEK1-Plk3 signaling cascade located to the Golgi membranes [22, 40, 41] (Figure 2).

GRASP55 and GRASP65 are two peripheral Golgi proteins localized to the *medial-trans* and *cis* cisternae membranes, respectively [42, 43]. Both GRASP proteins form homodimers which are able to maintain lateral cisternae connections and stacks by establishing *trans-*

oligomers with adjacent membranes homodimers in order to hold the Golgi ribbon-like structure (Figure 3A). In mitosis, *trans*-oligomers are inhibited by phosphorylation in their C-terminal serine/proline rich domain (SPR). Thus, when phosphorylation takes place, *trans*-oligomers are prevented (Figure 3A) and, as a consequence, the Golgi complex is disassembled [19, 27, 44-46]. Consistent with the role of GRASP proteins in Golgi G2 phase unlinking, antibodies against GRASP65, expression of GRASP domains mutants or non-regulatable mutants block mitotic Golgi fragmentation and delay mitotic entry [26, 27]. Similarly, microinjection of GRASP55 protein fragments or C-terminal mutants also leads to G2 block and decrease in mitotic index [25], suggesting that these proteins are both implicated in cisternae G2 phase unlinking in a cell cycle phosphorylation dependent manner. GRASP proteins are likely target candidates for regulatory signals. In mitosis, GRASP55 is phosphorylated by ERK2 downstream MEK1 [5, 20] whereas GRASP65 is phosphorylated by Plk1 [16, 18, 19] and Cdc2 [16, 19]. Besides, experimental evidences also suggest the existence of at least one additional kinase targeting GRASP65, since metabolically labeled cells arrested in mitosis by nocodazole show several phosphopeptides for Cdc2 and Plk1, but there is one additional phosphopeptide not accounted for them [16].

The sequential order of these phosphorylations is not yet known. These data indicate that during mitosis GRASP proteins are phosphorylated and, as a result, Golgi complex first unlinks and then unstacks thus facilitating vesicles and tubules formation. However, it is still unknown how Plk1 and Cdc2 are coordinated and organized at this stage of the process.

The membrane fission protein named CtBP3/BARS is involved in several membrane trafficking steps such as retrograde transport of KDEL receptor by COPI vesicles [47]. Its relevance in Golgi G2 unlinking phase has been demonstrated by its depletion, effect of dominant-negative mutants and antibodies against BARS experiments, all of which supports that BARS is required for G2/M transition and Golgi breakdown [4, 48]. Moreover, BARS is also phosphorylated in mitosis by an unknown kinase [49], all of these suggesting it might be a potential regulatory target in early stages of the process.

3. GOLGI STACKS DISASSEMBLY AND VESICULATION

From prophase to metaphase Golgi apparatus further breaks down into many smaller vesicles and tubules that accumulate around the spindle mitotic poles [50, 51]. Thereby, GRASP65 protein is phosphorylated so that *trans*-oligomerization and cisternal stacking are disrupted (Figure 3B) [19]. This unstacking facilitates COPI vesicle formation by increasing the amount of membrane surface, which is required for Golgi complex disassembly in mitosis [45]. Plk1 and Cdc2 are both involved in unstacking as they phosphorylate GRASP65 [16, 18, 19, 44], whereas GRASP55 is phosphorylated by ERK2 [5, 20]. Given that GRASP65 and GRASP55 are localized in different regions of the Golgi complex, it has been hypothesized that they play complementary functions in Golgi stacking and unstacking in a cell cycle phosphorylation dependent manner [25, 46, 52].

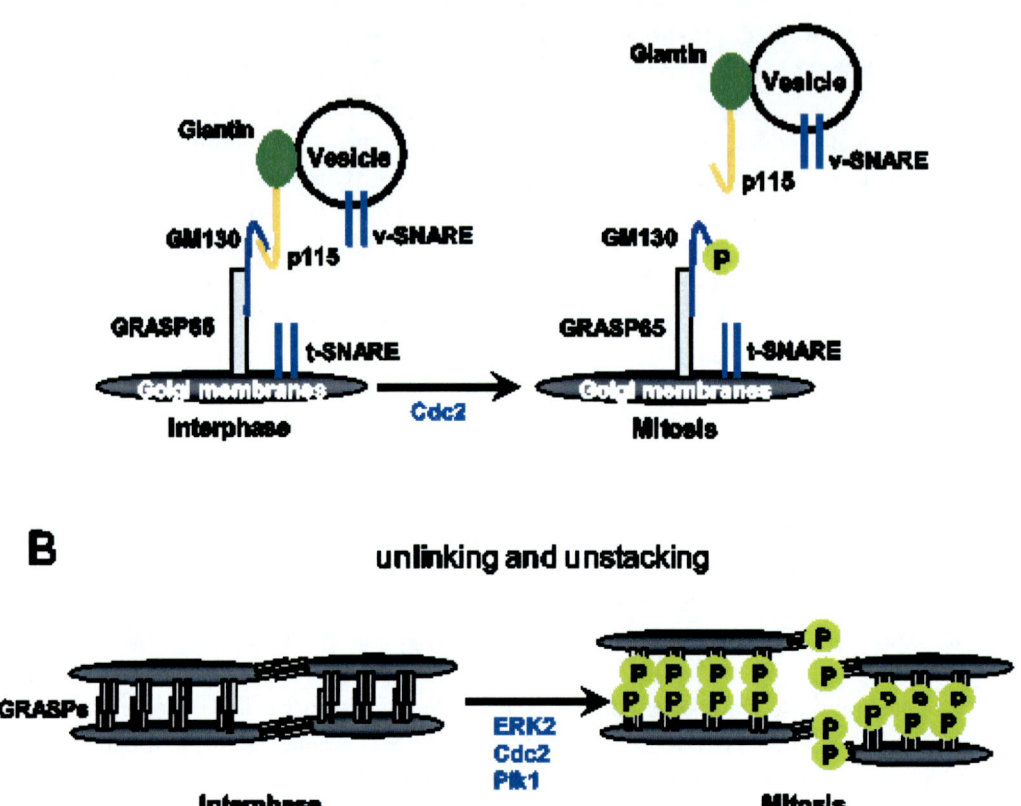

Figure 3. A. Cisternae lateral connections and stacks are held together by *trans*-oligomerization of GRASP proteins. In mitosis, GRASP proteins are phosphorylated so that the Golgi complex unlinks and unstacks. **B**. Vesicle fusion events. The initial tethering process is mediated by forming p115-GM130-Giantin complexes that are blocked by phosphorylation in mitosis.

Apart from unstacking, it is also required that vesicle budding remains active whereas membrane fusion complexes are inhibited, and thus smaller vesicles can be formed. The GTPase ARF-1 (Active ADP-ribosylation factor-1) is involved in recruitment of COPI protein complex implicated in membrane traffic [53]. In mitosis, some evidence indicates that ARF-1 remains active so that there is a continuous COPI vesicle formation and, as a consequence, fragmentation of Golgi membranes [24, 29, 54].

But this is still controversial since other authors have shown that ARF-1 is inactivated in mitosis [28, 31]. On the other hand, Cdc2 phosphorylates GM130 [15] and p47 [17] which results in the inactivation of NSF and VCP-dependent fusion pathways, respectively.

GM130 is a coiled-coil protein anchored by its C-terminus extreme to GRASP65 (Figure 3A) and thus to *cis*-Golgi stacks [42, 55, 56]. Besides, GM130 interacts by its N-terminus with the tethering factor p115 which mediates the initial vesicle tethering between COPI vesicles, through Giantin, and membranes, bridged by GM130, forming a p115-GM130-Giantin complex (Figure 3A) [56-58]. Next, it is formed the docking SNARE complex which defines the specificity of vesicle targeting [59, 60]. The role of NSF is to break the SNARE complex up, a process that needs ATP hydrolysis, and that results in vesicle fusion [61]. Cdc2 phosphorylates GM130 and prevents its binding with p115 and, therefore, it is blocked the initial vesicle tethering and also the fusion process dependent on NSF ATPase [15] (Figure 3A).

Less is known about VCP (Valosin-Containing Protein) mechanism of action in the fusion pathway. VCP is a member of the type II AAA ATPase family (ATPases associated with various cellular activities) that is implicated in a variety of cellular processes including membrane fusion, transcription activation, cell cycle control and apoptosis [62]. Similar to NSF, VCP interacts with the SNARE pair, in this case through the p47 cofactor, and break it up after ATP hydrolysis. Unlike NSF, VCP does not need the tethering process, but instead requires another cofactor named VCIP135 during the membrane fusion [60, 63, 64]. The phosphorylation of p47 by Cdc2 in mitosis inhibits its binding to the Golgi complex, thus the fusion is dependent on VCP [17].

In addition to reversible phosphorylation, it has been suggested a novel regulatory level that involves a cycle of ubiquitination and deubiquitination, which is also coordinated with Golgi membrane dynamics. In fact, the fusion pathway regulated by VCP is also controlled by an ubiquitination cycle; since two cofactors, p47 and VCIP135 are connected with ubiquitin modifications. p47 contains an ubiquitin binding motif (UBA) recognizing monoubiquitin, and VCIP135 is a deubiquitinating enzyme so that ubiquitinated proteins seem to be crucial in the VCP-mediated fusion mechanism [62, 65]. The attachment of a single ubiquitin, monoubiquitination, to the target protein serves as regulatory modification; in contrast to polyubiquitination, formed by several ubiquitin molecules that are attached into chains linked through Lys48 for proteasome degradation signaling, or Lys63 for DNA damage repair [66]. The fact is that experimental assays carried out with the ubiquitin mutant I44A support the notion of ubiquitination as a new Golgi dynamics controller. The I44A mutant acts as a dominant-negative that conjugates to target proteins but it is not recognized by VCP-p47 protein complex. Therefore, the ubiquitin mutant but not the wild type ubiquitin prevented cisternal regrowth in a reassembly assay since VCP-p47 complex is not able to recognize ubiquitinated Golgi proteins [67]; so that some unknown Golgi factors are required to be monoubiquitinated at the onset of mitosis in order to be later deubiquitinated in telophase. However, the possible targets and the exact mechanism by which ubiquitin signal functions in Golgi disassembly and reassembly are still poorly characterized [65, 67, 68].

4. GOLGI APPARATUS INHERITANCE AND THE MITOTIC SPINDLE

In metaphase, Golgi small vesicles are scattered throughout the cytoplasm (Figure 1) and they start to be distributed between daughter cells as the mitotic spindle is assembled. Two different models have been proposed trying to explain how this partitioning occurs. In one view, Golgi vesicles and tubules are absorbed into the endoplasmatic reticulum membranes and then reemerged from it, as it occurs with the nuclear envelope, which means that Golgi inheritance is mediated by the endoplasmatic reticulum [31, 69]. The second model suggests that Golgi fragments are partitioned independently of the endoplasmatic reticulum [6, 70-73], and proposes that the mitotic spindle is responsible for the partitioning [50, 51, 72]. There are some data supporting this second model. One is the observation of Golgi vesicles and tubules accumulated around the spindle poles [8, 50, 51, 74], whereas the endoplasmatic reticulum is excluded from that area [8, 71].

Recently, some experiments carried out by Wei and Seemann have finally demonstrated the link between the Golgi partitioning and the mitotic spindle since they have developed a procedure that induces an asymmetrical cell division so that one daughter cell (karyoplast) receives the entire spindle (centrosomes, chromosomes and spindle microtubules) and the other (cytoplast) lack all these [75]. Under these experimental conditions, the ribbon-like Golgi complex reforms in the karyoplasts whereas in the cytoplasts the Golgi stacks are scattered throughout their cytoplasm although they maintain their transport activity [76]. This has led to the proposal that there are two Golgi inheritance mechanisms: the spindle-independent by which elements of functional stacks are fragmented and distributed; and the spindle-dependent by which the ribbon determinants are partitioned between daughter cells [76]. The signals controlling the inheritance are not known but they are likely to be related to the interaction of vesicles with components of the centrosome and synchronized with cell cycle progression.

5. Golgi Apparatus Reassembly

In telophase, Golgi fragments start to reassemble by two mechanisms: membrane fusion processes mediated by two ATPases, VCP and NSF; and cisternae restacking mediated by dephosphorylation of GRASP proteins and GM130 (Figure 4). Vesicle fusion events dependent on NSF ATPase require two golgins, p115 and GM130, as well as SNARE pairs (GOS-28/Syntaxin-5). p115 is initially required for membrane regrowth and later for cisternal restacking through the formation of GM130-p115-Giantin complexes [30, 58, 60, 77]. The other ATPase implicated is VCP and its cofactors p47 [63], p37 [78] and VCIP135 [64], which form two different types of protein complexes: VCP-p47-VCIP135 and VCP-p37-VCIP135. In the first pathway, the protein complex that binds to Golgi membranes through Syntaxin-5 is disrupted after ATP hydrolysis [60, 64] and monoubiquitination has been demonstrated to be necessary for the process. As has been mentioned, monoubiquitination seems to be another level of regulation in Golgi dynamics in mitosis [67, 68]. In the second pathway, the protein complex formed by VCP-p37-VCIP135 uses GS-15 instead of Syntaxin-5 to bind Golgi membranes and, as in NSF pathway, it also requires p115 [78]. In contrast to NSF, deubiquitinating activity is not necessary in this case since p37 does not contain an ubiquitin binding motif (UBA) so that the complex VCP-p37 activity is not related with ubiquitination [78]. GM130 [23] and GRASP65 [24] dephosphorylation is mediated by PP2A (Bα). PP2A phosphatases are composed by a series of serine/treonine enzymes that functions as a trimeric complex formed by an invariable catalytic (C) and structural (A) subunits which bind to a variable regulatory one (B). Among these regulatory subunits, Bα is associated with the Golgi membranes and so that it functions specifically at the Golgi complex [23]. After membranes regrowth, cisternal restacking starts in order to reform the continuous and stacked Golgi membranous system characteristic of interphase. Restacking is initiated by p115 through its interaction with dephosphorylated GM130 and Giantin between adjacent membranes [58]. Besides, p115 is phosphorylated by CKII or CKII-like kinase that stimulates Giantin-GM130 binding, and stacking [79]. Finally, since GRASP65 homodimers are dephosphorylated by PP2A (Bα) [24], they are able to restore *trans*-oligomers with adjacent membranes homodimers and to establish lateral connections and Golgi stacks. Although there are not data about GRASP55 dephosphorylation, it is assumed that it is the case since GRASP65 and GRASP55 play complementary roles in Golgi cistenal stacking [46].

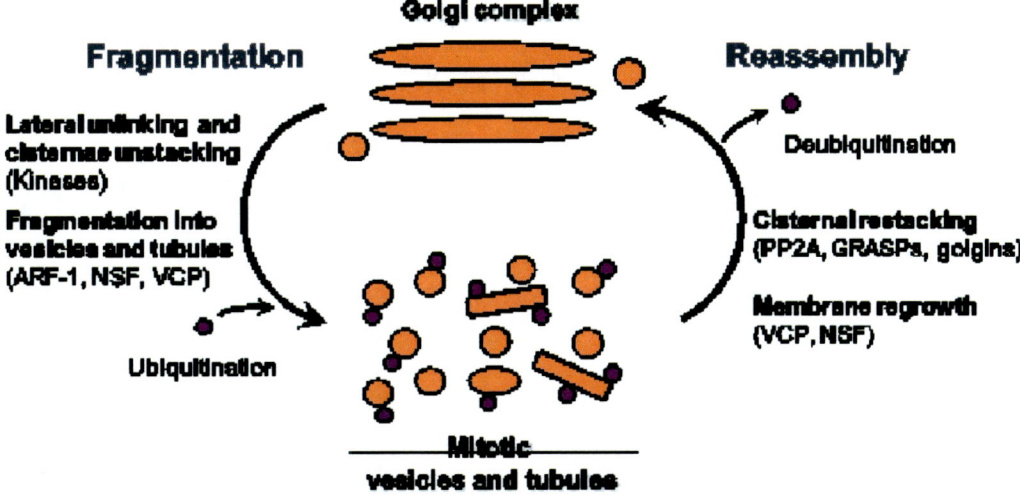

Figure 4. Diagram of the Golgi fragmentation, assembly and location where different signaling proteins participate.

CONCLUSION

Mitotic Golgi dynamics is a regulated and coordinated process that requires several molecular elements including kinases, structural Golgi proteins, small GTPases and ATPases (Figure 4). Phosphorylation is one of the main regulatory mechanisms of this process and although there have been described some kinases and phosphatases that are implicated, the interconnections between different signaling cascades is still not well characterized temporally or spatially. Therefore, the future challenge is to identify and characterize new signaling pathways implicated and identify kinases, their substrates and timing in mitosis, which will help to understand how the fragmentation and reassembly process is controlled.

ACKNOWLEDGMENTS

I.L-S. was supported by a fellowship from Ministerio de Educación, Ciencia e Innovación. Work in the laboratory was supported by grants from Ministerio de Educación, Ciencia e Innovación (SAF2007-60242, SAF2010-14935, and CSD2007-0017), Junta de Castilla y León (CSI14A08 and GR15).

REFERENCES

[1] Preuss D, Mulholland J, Franzusoff A, Segev N, Botstein D. Characterization of the Saccharomyces Golgi complex through the cell cycle by immunoelectron microscopy. *Mol. Biol. Cell*. 1992;3:789-803.

[2] Saint-Jore-Dupas C, Gomord V, Paris N. Protein localization in the plant Golgi apparatus and the trans-Golgi network. *Cell Mol. Life Sci*. 2004;61:159-71.

[3] Ladinsky MS, Mastronarde DN, McIntosh JR, Howell KE, Staehelin LA. Golgi structure in three dimensions: functional insights from the normal rat kidney cell. *J. Cell Biol*. 1999;144:1135-49.

[4] Colanzi A, Hidalgo Carcedo C, Persico A, Cericola C, Turacchio G, Bonazzi M, et al. The Golgi mitotic checkpoint is controlled by BARS-dependent fission of the Golgi ribbon into separate stacks in G2. *Embo J*. 2007;26:2465-76.

[5] Feinstein TN, Linstedt AD. Mitogen-activated protein kinase kinase 1-dependent Golgi unlinking occurs in G2 phase and promotes the G2/M cell cycle transition. *Mol. Biol. Cell*. 2007;18:594-604.

[6] Shorter J, Warren G. Golgi architecture and inheritance. *Annu. Rev. Cell Dev. Biol*. 2002;18:379-420.

[7] Colanzi A, Suetterlin C, Malhotra V. Cell-cycle-specific Golgi fragmentation: how and why? *Curr. Opin. Cell Biol*. 2003;15:462-7.

[8] Axelsson MA, Warren G. Rapid, endoplasmic reticulum-independent diffusion of the mitotic Golgi haze. *Mol. Biol. Cell*. 2004;15:1843-52.

[9] Persico A, Cervigni RI, Barretta ML, Colanzi A. Mitotic inheritance of the Golgi complex. *FEBS Lett*. 2009;583:3857-62.

[10] Colanzi A, Sutterlin C, Malhotra V. RAF1-activated MEK1 is found on the Golgi apparatus in late prophase and is required for Golgi complex fragmentation in mitosis. *J. Cell Biol*. 2003;161:27-32.

[11] Acharya U, Mallabiabarrena A, Acharya JK, Malhotra V. Signaling via mitogen-activated protein kinase kinase (MEK1) is required for Golgi fragmentation during mitosis. *Cell*. 1998;92:183-92.

[12] Colanzi A, Deerinck TJ, Ellisman MH, Malhotra V. A specific activation of the mitogen-activated protein kinase kinase 1 (MEK1) is required for Golgi fragmentation during mitosis. *J. Cell Biol*. 2000;149:331-9.

[13] Kano F, Takenaka K, Yamamoto A, Nagayama K, Nishida E, Murata M. MEK and Cdc2 kinase are sequentially required for Golgi disassembly in MDCK cells by the mitotic Xenopus extracts. *J. Cell Biol*. 2000;149:357-68.

[14] Shaul YD, Seger R. ERK1c regulates Golgi fragmentation during mitosis. *J. Cell Biol*. 2006;172:885-97.

[15] Lowe M, Rabouille C, Nakamura N, Watson R, Jackman M, Jamsa E, et al. Cdc2 kinase directly phosphorylates the cis-Golgi matrix protein GM130 and is required for Golgi fragmentation in mitosis. *Cell*. 1998;94:783-93.

[16] Lin CY, Madsen ML, Yarm FR, Jang YJ, Liu X, Erikson RL. Peripheral Golgi protein GRASP65 is a target of mitotic polo-like kinase (Plk) and Cdc2. *Proc. Natl. Acad. Sci. U S A*. 2000;97:12589-94.

[17] Uchiyama K, Jokitalo E, Lindman M, Jackman M, Kano F, Murata M, et al. The localization and phosphorylation of p47 are important for Golgi disassembly-assembly during the cell cycle. *J. Cell Biol.* 2003;161:1067-79.
[18] Sutterlin C, Lin CY, Feng Y, Ferris DK, Erikson RL, Malhotra V. Polo-like kinase is required for the fragmentation of pericentriolar Golgi stacks during mitosis. *Proc. Natl. Acad. Sci. U S A.* 2001;98:9128-32.
[19] Wang Y, Seemann J, Pypaert M, Shorter J, Warren G. A direct role for GRASP65 as a mitotically regulated Golgi stacking factor. *Embo J.* 2003;22:3279-90.
[20] Jesch SA, Lewis TS, Ahn NG, Linstedt AD. Mitotic phosphorylation of Golgi reassembly stacking protein 55 by mitogen-activated protein kinase ERK2. *Mol. Biol Cell.* 2001;12:1811-7.
[21] Xie S, Wang Q, Ruan Q, Liu T, Jhanwar-Uniyal M, Guan K, et al. MEK1-induced Golgi dynamics during cell cycle progression is partly mediated by Polo-like kinase-3. *Oncogene.* 2004;23:3822-9.
[22] Lopez-Sanchez I, Sanz-Garcia M, Lazo PA. Plk3 interacts with and specifically phosphorylates VRK1 in Ser342, a downstream target in a pathway that induces Golgi fragmentation. *Mol. Cell Biol.* 2009;29:1189-201.
[23] Lowe M, Gonatas NK, Warren G. The mitotic phosphorylation cycle of the cis-Golgi matrix protein GM130. *J. Cell Biol.* 2000;149:341-56.
[24] Tang D, Mar K, Warren G, Wang Y. Molecular mechanism of mitotic Golgi disassembly and reassembly revealed by a defined reconstitution assay. *J. Biol. Chem.* 2008;283:6085-94.
[25] Duran JM, Kinseth M, Bossard C, Rose DW, Polishchuk R, Wu CC, et al. The role of GRASP55 in Golgi fragmentation and entry of cells into mitosis. *Mol. Biol. Cell.* 2008;19:2579-87.
[26] Sutterlin C, Hsu P, Mallabiabarrena A, Malhotra V. Fragmentation and dispersal of the pericentriolar Golgi complex is required for entry into mitosis in mammalian cells. *Cell.* 2002;109:359-69.
[27] Tang D, Yuan H, Wang Y. The role of GRASP65 in Golgi cisternal stacking and cell cycle progression. *Traffic* (Copenhagen, Denmark). 2010;11:827-42.
[28] Altan-Bonnet N, Phair RD, Polishchuk RS, Weigert R, Lippincott-Schwartz J. A role for Arf1 in mitotic Golgi disassembly, chromosome segregation, and cytokinesis. *Proc. Natl. Acad. Sci. U S A.* 2003;100:13314-9.
[29] Xiang Y, Seemann J, Bisel B, Punthambaker S, Wang Y. Active ADP-ribosylation factor-1 (ARF1) is required for mitotic Golgi fragmentation. *J. Biol. Chem.* 2007;282:21829-37.
[30] Rabouille C, Levine TP, Peters JM, Warren G. An NSF-like ATPase, p97, and NSF mediate cisternal regrowth from mitotic Golgi fragments. *Cell.* 1995;82:905-14.
[31] Altan-Bonnet N, Sougrat R, Liu W, Snapp EL, Ward T, Lippincott-Schwartz J. Golgi inheritance in mammalian cells is mediated through endoplasmic reticulum export activities. *Mol. Biol. Cell.* 2006;17:990-1005.
[32] Colanzi A, Corda D. Mitosis controls the Golgi and the Golgi controls mitosis. *Curr. Opin. Cell Biol.* 2007;19:386-93.
[33] Chang L, Karin M. Mammalian MAP kinase signalling cascades. *Nature.* 2001;410:37-40.

[34] Shapiro PS, Vaisberg E, Hunt AJ, Tolwinski NS, Whalen AM, McIntosh JR, et al. Activation of the MKK/ERK pathway during somatic cell mitosis: direct interactions of active ERK with kinetochores and regulation of the mitotic 3F3/2 phosphoantigen. *J. Cell Biol.* 1998;142:1533-45.

[35] Dangi S, Shapiro P. Cdc2-mediated inhibition of epidermal growth factor activation of the extracellular signal-regulated kinase pathway during mitosis. *J. Biol. Chem.* 2005;280:24524-31.

[36] Laird AD, Morrison DK, Shalloway D. Characterization of Raf-1 activation in mitosis. *J. Biol. Chem.* 1999;274:4430-9.

[37] Aebersold DM, Shaul YD, Yung Y, Yarom N, Yao Z, Hanoch T, et al. Extracellular signal-regulated kinase 1c (ERK1c), a novel 42-kilodalton ERK, demonstrates unique modes of regulation, localization, and function. *Mol. Cell Biol.* 2004;24:10000-15.

[38] Lippincott-Schwartz J, Yuan LC, Bonifacino JS, Klausner RD. Rapid redistribution of Golgi proteins into the ER in cells treated with brefeldin A: evidence for membrane cycling from Golgi to ER. *Cell.* 1989;56:801-13.

[39] Seemann J, Jokitalo E, Pypaert M, Warren G. Matrix proteins can generate the higher order architecture of the Golgi apparatus. *Nature.* 2000;407:1022-6.

[40] Ruan Q, Wang Q, Xie S, Fang Y, Darzynkiewicz Z, Guan K, et al. Polo-like kinase 3 is Golgi localized and involved in regulating Golgi fragmentation during the cell cycle. *Exp. Cell Res.* 2004;294:51-9.

[41] Valbuena A, Lopez-Sanchez I, Vega FM, Sevilla A, Sanz-Garcia M, Blanco S, et al. Identification of a dominant epitope in human vaccinia-related kinase 1 (VRK1) and detection of different intracellular subpopulations. *Arch. Biochem. Biophys.* 2007;465:219-26.

[42] Barr FA, Puype M, Vandekerckhove J, Warren G. GRASP65, a protein involved in the stacking of Golgi cisternae. *Cell.* 1997;91:253-62.

[43] Shorter J, Watson R, Giannakou ME, Clarke M, Warren G, Barr FA. GRASP55, a second mammalian GRASP protein involved in the stacking of Golgi cisternae in a cell-free system. *Embo J.* 1999;18:4949-60.

[44] Wang Y, Satoh A, Warren G. Mapping the functional domains of the Golgi stacking factor GRASP65. *J. Biol. Chem.* 2005;280:4921-8.

[45] Wang Y, Wei JH, Bisel B, Tang D, Seemann J. Golgi cisternal unstacking stimulates COPI vesicle budding and protein transport. *PLoS ONE.* 2008;3:e1647.

[46] Xiang Y, Wang Y. GRASP55 and GRASP65 play complementary and essential roles in Golgi cisternal stacking. *J. Cell Biol.* 2010;188:237-51.

[47] Yang JS, Lee SY, Spano S, Gad H, Zhang L, Nie Z, et al. A role for BARS at the fission step of COPI vesicle formation from Golgi membrane. *Embo J.* 2005;24:4133-43.

[48] Hidalgo Carcedo C, Bonazzi M, Spano S, Turacchio G, Colanzi A, Luini A, et al. Mitotic Golgi partitioning is driven by the membrane-fissioning protein CtBP3/BARS. *Science* (New York, NY. 2004;305:93-6.

[49] Boyd JM, Subramanian T, Schaeper U, La Regina M, Bayley S, Chinnadurai G. A region in the C-terminus of adenovirus 2/5 E1a protein is required for association with a cellular phosphoprotein and important for the negative modulation of T24-ras mediated transformation, tumorigenesis and metastasis. *Embo J.* 1993;12:469-78.

[50] Shima DT, Cabrera-Poch N, Pepperkok R, Warren G. An ordered inheritance strategy for the Golgi apparatus: visualization of mitotic disassembly reveals a role for the mitotic spindle. *J. Cell Biol*. 1998;141:955-66.

[51] Wei JH, Seemann J. Spindle-dependent partitioning of the Golgi ribbon. *Commun. Integr. Biol*. 2009;2:406-7.

[52] Feinstein TN, Linstedt AD. GRASP55 regulates Golgi ribbon formation. *Mol. Biol. Cell*. 2008;19:2696-707.

[53] Rothman J. Mechanisms of intracellular protein transport. *Nature*. 1994;372:55-63.

[54] Misteli T, Warren G. COP-coated vesicles are involved in the mitotic fragmentation of Golgi stacks in a cell-free system. *J. Cell Biol*. 1994;125:269-82.

[55] Barr FA, Nakamura N, Warren G. Mapping the interaction between GRASP65 and GM130, components of a protein complex involved in the stacking of Golgi cisternae. *Embo J*. 1998;17:3258-68.

[56] Nakamura N, Rabouille C, Watson R, Nilsson T, Hui N, Slusarewicz P, et al. Characterization of a cis-Golgi matrix protein, GM130. *J. Cell Biol*. 1995;131:1715-26.

[57] Nakamura N, Lowe M, Levine TP, Rabouille C, Warren G. The vesicle docking protein p115 binds GM130, a cis-Golgi matrix protein, in a mitotically regulated manner. *Cell*. 1997;89:445-55.

[58] Sonnichsen B, Lowe M, Levine T, Jamsa E, Dirac-Svejstrup B, Warren G. A role for giantin in docking COPI vesicles to Golgi membranes. *J. Cell Biol*. 1998;140:1013-21.

[59] Rothman JE, Warren G. Implications of the SNARE hypothesis for intracellular membrane topology and dynamics. *Curr. Biol*. 1994;4:220-33.

[60] Rabouille C, Kondo H, Newman R, Hui N, Freemont P, Warren G. Syntaxin 5 is a common component of the NSF- and p97-mediated reassembly pathways of Golgi cisternae from mitotic Golgi fragments in vitro. *Cell*. 1998;92:603-10.

[61] Sollner T, Whiteheart SW, Brunner M, Erdjument-Bromage H, Geromanos S, Tempst P, et al. SNAP receptors implicated in vesicle targeting and fusion. *Nature*. 1993;362:318-24.

[62] Wang Q, Song C, Li CC. Molecular perspectives on p97-VCP: progress in understanding its structure and diverse biological functions. *J. Struct. Biol*. 2004;146:44-57.

[63] Kondo H, Rabouille C, Newman R, Levine TP, Pappin D, Freemont P, et al. p47 is a cofactor for p97-mediated membrane fusion. *Nature*. 1997;388:75-8.

[64] Uchiyama K, Jokitalo E, Kano F, Murata M, Zhang X, Canas B, et al. VCIP135, a novel essential factor for p97/p47-mediated membrane fusion, is required for Golgi and ER assembly in vivo. *J. Cell Biol*. 2002;159:855-66.

[65] Meyer HH. Golgi reassembly after mitosis: the AAA family meets the ubiquitin family. *Biochim. Biophys. Acta*. 2005;1744:481-92.

[66] Schnell JD, Hicke L. Non-traditional functions of ubiquitin and ubiquitin-binding proteins. *J. Biol. Chem*. 2003;278:35857-60.

[67] Wang Y, Satoh A, Warren G, Meyer HH. VCIP135 acts as a deubiquitinating enzyme during p97-p47-mediated reassembly of mitotic Golgi fragments. *J. Cell Biol*. 2004;164:973-8.

[68] Meyer HH, Wang Y, Warren G. Direct binding of ubiquitin conjugates by the mammalian p97 adaptor complexes, p47 and Ufd1-Npl4. *Embo J*. 2002;21:5645-52.

[69] Zaal KJ, Smith CL, Polishchuk RS, Altan N, Cole NB, Ellenberg J, et al. Golgi membranes are absorbed into and reemerge from the ER during mitosis. *Cell.* 1999;99:589-601.
[70] Pelletier L, Jokitalo E, Warren G. The effect of Golgi depletion on exocytic transport. *Nat. Cell Biol.* 2000;2:840-6.
[71] Jesch SA, Mehta AJ, Velliste M, Murphy RF, Linstedt AD. Mitotic Golgi is in a dynamic equilibrium between clustered and free vesicles independent of the ER. *Traffic* (Copenhagen, Denmark). 2001;2:873-84.
[72] Jokitalo E, Cabrera-Poch N, Warren G, Shima DT. Golgi clusters and vesicles mediate mitotic inheritance independently of the endoplasmic reticulum. *J. Cell Biol.* 2001;154:317-30.
[73] Barr FA. Golgi inheritance: shaken but not stirred. *J. Cell Biol.* 2004;164:955-8.
[74] Seemann J, Pypaert M, Taguchi T, Malsam J, Warren G. Partitioning of the matrix fraction of the Golgi apparatus during mitosis in animal cells. *Science* (New York, NY. 2002;295:848-51.
[75] Wei JH, Seemann J. Induction of asymmetrical cell division to analyze spindle-dependent organelle partitioning using correlative microscopy techniques. *Nature protocols.* 2009;4:1653-62.
[76] Wei JH, Seemann J. The mitotic spindle mediates inheritance of the Golgi ribbon structure. *J. Cell Biol.* 2009;184:391-7.
[77] Shorter J, Warren G. A role for the vesicle tethering protein, p115, in the post-mitotic stacking of reassembling Golgi cisternae in a cell-free system. *J. Cell Biol.* 1999;146:57-70.
[78] Uchiyama K, Totsukawa G, Puhka M, Kaneko Y, Jokitalo E, Dreveny I, et al. p37 is a p97 adaptor required for Golgi and ER biogenesis in interphase and at the end of mitosis. *Dev Cell.* 2006;11:803-16.
[79] Dirac-Svejstrup AB, Shorter J, Waters MG, Warren G. Phosphorylation of the vesicle-tethering protein p115 by a casein kinase II-like enzyme is required for Golgi reassembly from isolated mitotic fragments. *J. Cell Biol.* 2000;150:475-88.
[80] Yoshimura S, Yoshioka K, Barr FA, Lowe M, Nakayama K, Ohkuma S, et al. Convergence of cell cycle regulation and growth factor signals on GRASP65. *J. Biol. Chem.* 2005;280:23048-56.

Chapter 4

GOLGI APPARATUS AND HYPERICIN-MEDIATED PHOTODYNAMIC ACTION

*Chuanshan Xu[1,2], Xinshu Xia[2], Hongwei Zhang[1] and Albert Wingnang Leung[1]**

[1]Department of Rehabilization Medicine, The First Affiliated Hospital, Chongqing Medical University, Chongqing, China
[2]School of Chinese Medicine, The Chinese University of Hong Kong, Shatin, Hong Kong, China

ABSTRACT

Hypericin isolated from *Hypericum perforatum* plants, exhibits a wide range of biological activities and medical applications for treating tumors. Emerging evidence has demonstrated that hypericin could be activated by visible light to produce reactive oxygen species (ROS) which destroys a tumor directly or indirectly. Hypercin-induced photodynamic therapy has showed considerable promise as an alternative modality in the management of malignant tumors. However, the exact mechanisms need to be clarified. Recent studies have showed that hypericin was accumulated in the Golgi apparatus, indicating that the role of the Golgi apparatus is indispensable in the biological mechanisms of hypericin-mediated photodynamic therapy.

Keywords: hypericin; photodynamic therapy; Golgi apparatus; tumor

Malignant tumors are one of the most dangerous diseases threatening human life. The common modalities classically used in the tumor therapy are mainly surgery, radiotherapy, and chemotherapy. Although these therapeutic methods can improve the clinical outcomes, serious side-effects, high recurrence, and multi-drug resistance have become the main limitations of these classical modalities in the clinical application [1, 2]. Therefore, the exploration of novel and effective therapeutic strategies is being pursued.

---

* School of Chinese Medicine, The Chinese University of Hong Kong, Shatin, Hong Kong, China. E-mail: awnleung@cuhk.edu.hk

1. AN ALTERNATIVE THERAPEUTIC MODALITY: PHOTODYNAMIC THERAPY

The application of light in the disease treatment has several thousand years of history. Photodynamic therapy (PDT) is the apotheosis of light as a therapeutic method in the clinical application. About 4000 years ago, the Egyptians began to use light to treat vitiligo with a plant containing photosensitizer. In 1900, a German Raal found that light-irradiated acridine orange could dramatically kill paramecium. The first clinical application of PDT on a tumor was reported by von Tappeiner and Jesionek in 1903 [3, 4, 5]. PDT can be developed to become an alternative therapy for tumors; the credit should be given to Dougherty. In 1975, Dougherty et al. first reported that red light-activated hematoporphyrin derivative (HpD) completely destroyed mouse mammary tumors. Subsequent clinical trials using HpD-PDT to treat bladder and skin tumors achieved limited but exciting success [5]. To date, accumulating experimental studies and clinical trials have shown that PDT can effectively eradicate tumors that light can reach [6, 7, 8]. The efficacy of PDT depends on the generation of reactive oxygen species (ROS) produced by light-activated photosensitizers which preferentially retain in tumor tissues [9, 10, 11, 12]. In the presence of oxygen molecules, the light-activated photosensitizer inside tumor tissues can cause a series of photochemical reactions and consequently the generation of cytotoxic ROS, which can directly and/or indirectly destroy tumor tissues [13, 14]. In the photochemical reaction, PDT involves the interaction of photosensitizers, light and oxygen, mainly including type I and type II photochemical reactions [15]. Following the absorption of appropriate photons, the photosensitizer is excited into a higher energy state (an excited singlet state or excited triplet state). When the activated photosensitizer returns to ground state, an electron is often transferred to adjacent molecules to form free radicals, namely a type I photochemical reaction. The free radicals then interact rapidly with molecular oxygen to produce ROS, which include the superoxide anion radical, hydrogen peroxides, and hydroxyl radicals. Alternatively, when the energy is directly transferred to ground-state molecular oxygen, type II photochemical reaction is produced. In the type II reactions, the direct interaction of the activated photosensitizer with molecular oxygen produces highly reactive singlet oxygen ($^1O_2$). The singlet oxygen initiates further oxidative reactions with proximate lipid membranes, enzymes, and nucleic acids to result in the oxidative lethal damage of tumor cells [16, 17, 18]. Compared to the classical modalities of malignancies, PDT shows many advantages: (1) cost-effectiveness and higher cure rate, (2) low side-effects, (3) less complications, (4) higher life quality, and (5) diagnosis accompanying therapy [5, 19, 20]. Therefore, PDT has been regarded as a promising alternative for treating malignancies. The clinical application of PDT with photofrin has been approved for treating esophagus cancer in USA, Canada, France, and Japan, for lung cancer in Japan, Holland and Germany, for bladder cancer in Canada, and for gastric and cervical cancers in Japan [21]. The successful application of PDT in tumor therapy depends on the photosensitizer and its matched light source. The light sources currently used are quartz-tungsten-halogen bulbs and Laser. The quartz-tungsten-halogen bulb has lower conversion efficiency of electric energy to light and higher heat production.

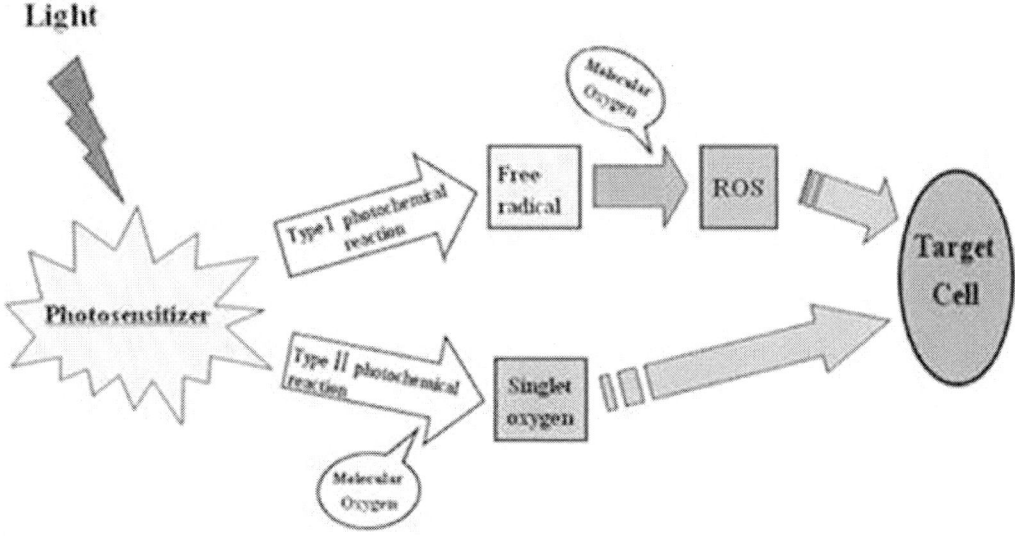

Figure 1. Photodynamic reaction depends on the presence of Molecular oxygen.

Laser is often expensive and has a short lifespan. This limits the application of quartz-tungsten-halogen bulbs and Laser in PDT as light source. The recently developed light emitting diode (LED) has been proposed as an alternative light source for PDT because LED is inexpensive, has a small volume, longer lifespan, and their light intensity can be adjustable [22, 23]. Moreover, LED can be arranged in arrays to irradiate larger areas. A novel LED light source has been setup in our lab [24, 25]. The wavelengths of the light source from LED are 630 nm and 590nm, and the highest output power reaches 130 mW/cm^2 in the area of 78.5 cm^2.

2. A Promising Second-Generation Photosensitive Drug: Hypericin

The photosensitizer is a key component affecting the PDT efficacy. The first-generation photosensitizer is hematoporphyrin derivative (HpD), for example, Photofrin. Although Photofrin has been approved in many countries for cancers locating in the lung, digestive tract and genitourinary tract [26, 27], Photofrin is a complex mixture of HpD, which has very poor absorption of light at long wavelength and prolonged skin photosensitivity. These drawbacks limit the clinical application of the first-generation photosensitizer [28, 29, 30]. The second-generation photosensitizers are mainly from synthetic chemicals with unique composition, strong absorption of light at long wavelength and short skin retention [30]. The photosensitizers currently include benzoporphyrins, pheophorbides, phthalocyanines, methylene blue derivative, and 5-ALA [27, 31, 32, 33, 34]. Recently, several extracts from traditional Chinese herbs have been found to be photosensitive, for example, curcumin from the *Curcuma longa* (Figure 2) and hypocrellins from *Hypocrella bambuase* (Figure 3) [35, 36]. *Hypericum perforatum* (Figure 4), also known as *St. John's Wort* in Western medicine, is an indigenous herb in Eastern China, Western and Eastern Sichuan, Gueizhou, and Xinjiang

[37, 38]. In traditional Chinese Medicine, *Hypericum perforatum* is bitter, sweet and cold, and categorized as a heat-clearing medicinal, which clears heat and resolves toxins. Modern pharmacological studies have shown that the medicinal displays various biological activities including anti-inflammatory, antiviral, antidepressant, and anti-tumor activities [39, 40, 41, 42]. The water extract of Hypericum perforatum has been used a remedy against urogenital inflammations, diabetes mellitus and heart diseases.

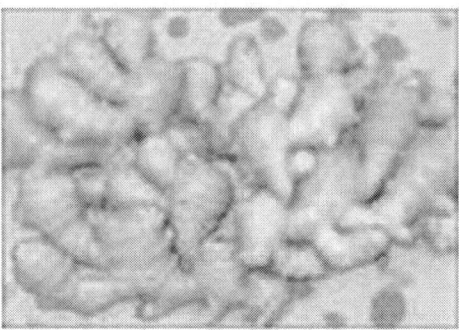

Figure 2. Curcuma longa (ggene.cn/.../NLKX/nyzc/2009/0820/2644.html).

Figure 3. Hypocrella bambuase (www.zgycsc.com/salebuymc.php?tp=1&mcid=26...).

Figure 4. Hypericum perforatum http://hi.baidu.com/leiazarias/blog/item/fb744a111 a0e49c2a6ef3feb .html).

The ethanol extract of the medicinal exhibits potent antiulcer activity and Süntar et al. proved that the aerial parts of hypericum perforatum possess significant wound healing and antiflammatory activities [43]. Hypericin (Figure 5), a hydroxylated phenanthroperylenequinone isolated from *Hypericum perforatum* plants, displays a wide range of biological activities and medicinal applications for treating depression, tumors and infectious diseases [39, 44, 45]. Emerging evidences demonstrates that hypericin is photosensitive. In ancient times, animals that ingested hypericin-containing plants were found to have hypericism, and be severely sensitive to light [46]. Upon light illumination, hypericin can change the energy state to produce highly reactive oxygen species such as singlet oxygen, superoxide anion and hydrogen peroxide, which can damage the proximate cells and tissues. Light-activated hypericin can significantly kill tumor cells, including human glioma cells, nasopharyngeal carcinoma cells, pituitary tumor cells, and bladder cancer cells [47, 48, 49, 50]. Interestingly, hypericin is very safe and does not show genotoxicity, or toxicity in humans and animals [39, 49, 51]. Therefore, hypericin has recently been put forward as a new promising "second-generation" photosensitizer for treating tumors. However, the biological mechanisms of hypericin-mediated PDT remain unclear.

3. THE MECHANISMS OF HYPERICIN-MEDIATED PDT AND GOLGI APPARATUS'S ROLE

Hypericin-mediated PDT induces cell death through apoptosis or necrosis depending on the degree of external damages [19, 52, 53]. A cell often undergoes necrosis when the energy in the cell is not enough to execute the apoptotic process. Apoptosis has been proven to be the main mode of cell death induced by hypericin-mediated PDT. Morphological and biochemical changes can distinguish apoptotic cells from necrosis [54, 55]. Ali et al reported that cell shrinkage, sub-diploid DNA increase and phosphatidylserine (PS) externalization have occurred in the CNE2 cells treated with hypericin-induced PDT, indicating the hypericin-induced apoptosis after photosensitization of hypericin [56]. An apoptotic body such as nuclear condensation and fragmentation in CNE2 cells were observed eight hours after photodynamic therapy with hypericin. Moreover, flow cytometry with annexin-FITC/propidium iodide (PI) staining showed that early and late apoptotic (necrotic) rate significantly increased, in which the early apoptotic rate significantly increased up to 53.08% and the late apoptotic (necrotic) rate only increased to 6.77% [39]. These demonstrated that apoptosis induction was one of the prominent mechanisms in hypericin-PDT.

Cell apoptosis occurs through two major pathways including the mitochondria-dependent and the death receptor-dependent pathways. Mitochondria, as critical executers, have been proven to play a central role in the regulation of apoptosis [57, 58]. The mitochondrial damages often interfere with cellular breathing, energy metabolism, and even make the permeability transition pore (PTP) open which leads to the collapse of the mitochondria membrane potential (MMP). The PTP open and MMP collapse cause the release of mitochondrial cytochrome c into cytoplasm and initiate Caspase cascades, where the cells will undergo apoptosis [2, 59, 60]. Chaloupka et al reported that hypericin photoxicity-induced cell apoptosis was triggered by photon-induced mitochondrial function alteration [61]. Recently, several secondary mitochondrial events have been found to be associated with hypericin-induced phototoxicity in cells. A mitochondrial apoptotic cascade through the activation of caspases, release of cytochrome c into the cytosol and decrease of mitochondrial membrane potential was observed after photodynaminc treatment with hypericin. In subsequent studies, hypericin-mediated PDT induced an intracellular pH decrease in glioma cells and inhibition of mitochondria hexokinase bound on the outer membrane, demonstrating that mitochondria were a target of hypericin phototoxicity, and the mitochondria-dependent pathway was involved in apoptosis induced by PDT with hypericin [62]. The death receptor-dependent apoptotic pathway involves the CD95 (Fas) death receptor. The CD95 death receptor activated by Fas ligand results in the formation of a death-inducing signaling complex (DISC). Apart from the receptor-ligand complex, the recruitment of caspase-8 to DISC also initiates its proteolytic activation and induces the subsequent activation of downstream caspases [63, 64, 65]. Recently, Ali et al reported that hypericin-mediated PDT caused apoptosis of tumor cells through death receptor-dependent pathway [66], indicating that the death receptor-dependent pathway might be the other mechanism of cell death.

In PDT-induced apoptosis, reactive oxygen species (ROS) is an important initiator [52]. As ROS has a very short lifespan and limited diffusion distance in biological tissues, the ROS-induced damages always happens close to the localization of the photosensitizer in the biological tissues [17]. Therefore, there is a strong relationship between the intracellular organelles, where the photosensitizer locates, and the apoptotic signaling pathway [27]. The roles of some intracellular organelles, such as lysosome, mitochodondria, and endoplasmic reticulum, have been focused on in the PDT-induced apoptotic signaling pathways. However, the Golgi apparatus has been overlooked. Its role in the PDT needs to be clarified.

Since 1898, Camillo Golgi found and described the diffuse reticular network of the Golgi apparatus in the cells, but it was not until a half-century later that the ribbon like cisternal architecture was observed using an electron microscope. In normal cells, the Golgi complex is factually a series of flattened, parallel, interconnected cisternae located in the perinuclear region. The functions of the Golgi apparatus involve the transport, the processing, and the storage of proteins synthesized in the rough endoplasmic reticulum and destined to be secreted [67, 68]. Various apoptosis-associated molecules, for example, caspase-2 and -8, are found to localize in the Golgi complex. Caspase-2 is a unique caspase with the functions of both the initiator and the effector [69]. As an initiator caspase, the activation of caspase-2 in response to apoptotic stresses can open up the pores of the mitochondrial membrane and trigger the release of cytochrome c from mitochondria, regulate the activities of effector caspases (e.g., caspase-3, -7) and induce the generation of tBid to induce apoptosis. Caspase-2 functions as an effector caspase similar to the executioner caspase-3 and caspase-7, which directly cause apoptosis [70]. Caspase-8 is also an initiator caspase responsible for cleavage and activation of the effector caspases. Caspase-8 is activated when bound to DISC [71] and it subsequently causes the effector caspase activation to cause cell apoptosis through the death receptor-dependent pathway [63, 64]. Additionally, Golgi apparatus is also found to be associated with genglioside GD3 synthesis. The ceramide is often converted by GD3 synthase to GD3. GD3 can shuttle mitochondria, and subsequently trigger mitochondrial membrane permeabilization, and induce apoptosis [69]. Recently, Ferri and Kroemer found that the Golgi apparatus-specific apoptotic pathway existed in various biological systems. Brefeldin A, an agent disrupting the Golgi apparatus, was confirmed to be an effective apoptosis inducer through mitochondrial dysfunction and caspase activation [72]. Moreover, effective inhibition of Golgi fragmentation and dispersal can decrease or delay cell apoptosis, suggesting the crucial role of Golgi apparatus in cell apoptosis [68]. Interestingly, Matroule et al. found that photosensitizer pyropheophorbide-a methyester, could accumulate in several intracellular organelles such as ER, lysosomes and the Golgi complex, and the photosensitizer-mediated PDT-induced damage on the Golgi apparatus were very important in the induction of apoptotic cell death. These findings indicated that the Golgi apparatus was a novel target of PDT in tumor cells [73]. In order to clarify the role of the Golgi apparatus in PDT-induced apoptosis, Ogata et al. used 2,4,5,7-tetrabromorhodamine 123 bromide (TBR) which only incorporates into the Golgi apparatus as a photosensitizer. Their results showed that TBR-mediated PDT effectively induced apoptosis of HeLa cells and Golgi dysfunction played an important role in apoptotic signaling triggering by TBR-induced PDT [27].

Golgi apparatus is a target organelle where hypericin preferentially accumulates [62, 74, 75]. Upon light radiation, the production of highly reactive oxygen species from the photosensitization of hypericin on the membrane of the Golgi apparatus causes the breakage of the Golgi membrane to release the apoptosis-associated molecules, for example, caspase-2, caspase-8, and so on. These apoptosis-associated molecules activate the downstream effector caspases in the cytosol or act directly as an effector caspase to initiate the mitochondria-dependent or death receptor-dependent pathway to induce apoptosis of the treated cells. However, no direct evidence has been reported on the Golgi apparatus' role in the bioeffect of hypericin-induced PDT. In view of the abovementioned facts from our perspective, the Golgi apparatus is an indispensable organelle for clarifying the mechanisms of hypericin-mediated PDT on tumors. Therefore, the exact mechanism should be determined in our following investigations.

ACKNOWLEDGMENTS

This work was supported by the grants from National Nature Science Foundation of China (30973168) and the Chinese University of Hong Kong (2030391, 2030408). We express our sincere thanks to Professor Tinghe Yu and Mr. Wah Lap Cheung for their helpful assistance.

REFERENCES

[1] Lin YT, Wang LF, Hsu YC. Curcuminoids Suppress the Growth of Pharynx and Nasopharyngeal Carcinoma Cells through Induced Apoptosis. *J. Agric. Food Chem.* 2009;57:3765-70.

[2] Wang P, Xu CS, Xu J, Wang X, Leung AW. Hypocrellin B enhances ultrasound-induced cell death of nasopharyngeal carcinoma cells. *Ultrasound in medicine and biology*. 2010;36:336-342.

[3] Wang T, Ma J. As the light took effect, the symptoms vanished-introduction of photodynamic therapy. *Ziran Zazhi*.2000;22:276-281.

[4] Triesscheijn, P. Baas, J.H. Schellens and F.A. Stewart, Photodynamic therapy in oncology, *Oncologist* 2006;11:1034–1044.

[5] Chatterjee DK, Fong LS, Zhang Y. Nanoparticles in photodynamic therapy: an emerging paradigm. *Advanced drug delivery reviews*. 2008,60:1627-1637.
[6] Lorenz KJ, Maier H. Photodynamic therapy with meta-tetrahydroxyphenylchlorin (Foscan) in the management of squamous cell carcinoma of the head and neck: experience with 35 patients. *Eur. Arch. Otorhinolaryngol*. 2009;266(12):1937-44.
[7] Rees JR, Lao-Sirieix P, Wong A, Fitzgerald RC. Treatment for Barrett's oesophagus. *Cochrane Database Syst. Rev*. 2010;1:CD004060.
[8] Allison RR, Sibata CH. Oncologic photodynamic therapy photosensitizers: a clinical review. *Photodiagnosis Photodyn. Ther*. 2010;7:61-75.
[9] Mak NK, Kok TW, Wong RN, Lam SW, Lau YK, Leung WN, Cheung NH, Huang DP, Yeung LL, Chang CK. Photodynamic activities of sulfonamide derivatives of porphycene on nasopharyngeal carcinoma cells. *J. Biomedl. Sci*. 2003;10:418-429.
[10] Leung W.N., Sun X., Mak N.K., Yow C.M. Photodynamic effects of mTHPC on human colon adenocarcinoma cells: photocytotoxicity, subcellular localiztion and apoptosis. *Photochem. Photobiol*. 2002; 75: 406-411.
[11] Xu CS and Leung AWN. Photodynamic Effects of Pyropheophorbids-a Methyl Ester in Nasopharyngeal Carcinoma Cells. *Med. Sci. Monit*, 2006; 12: BR257-262.
[12] Yow CM, Chen JY, Mak NK, Cheung NH, Leung AWN. Cellular uptake, subcellular localization and photodamaging effect of temoporfin (mTHPC) in nasopharyngeal carcinoma cells: comparison with hematoporphyrin derivative. *Cancer Lett*. 2000; 157: 123-31.
[13] Yow CM, Wong CK, Huang Z, Ho RJ. Study of the efficacy and mechanism of ALA-mediated photodynamic therapy on human hepatocellular carcinoma cell. *Liver Int*. 2007; 27:201-208.
[14] Wei Y, Xing D, Luo S, Yang L, Chen Q. Quantitative measurement of reactive oxygen species in vivo by utilizing a novel method: chemiluminescence with an internal fluorescence as reference. *J. Biomed. Opt*. 2010;15:027006.
[15] Kamuhabwa AR, Agostinis PM, D'Hallewin MA, Baert L, de Witte PA. Cellular photodestruction induced by hypericin in AY-27 rat bladder carcinoma cells. *Photochem. Photobiol*. 2001;74:126-32.
[16] Moan J, Peng Q, Sorensen R, Jani V, Nesland JM. Biophysical foundations of photodynamic therapy. *Endoscopy*.1998;30:387-391.
[17] Buytaert E, Dewaele M, Agostinis P. Molecular effectors of multiple cell death pathways initiated by photodynamic therapy. *Biochimica et Biophysica Acta*. 2007,1776:86-107.
[18] Alves E, Carvalho CM, Tomé JP, Faustino MA, Neves MG, Tomé AC, Cavaleiro JA, Cunha A, Mendo S, Almeida A. Photodynamic inactivation of recombinant bioluminescent Escherichia coli by cationic porphyrins under artificial and solar irradiation. *J. Ind. Microbio Biotechnol*. 2008;35:1447-54.
[19] Kleban J, Mikes J, Horváth V, Sacková V, Hofmanová J, Kozubík A, Fedorocko P. Mechanisms involved in the cell cycle and apoptosis of HT-29 cells pre-treated with MK-886 prior to photodynamic therapy with hypericin. *Journal of Photochemistry and Photobiology B:Biology*. 2008,93:108-118.
[20] Liu JY, Jiang XJ, Fong WP, Ng DK. Synthesis, Characterization, and In Vitro Photodynamic Activity of Novel Amphiphilic Zinc(II) Phthalocyanines Bearing Oxyethylene-Rich Substituents. *Met Based Drugs*. 2008;2008:284691-8.

[21] Ma JS. Development of second generational porphyrin-based photosensitizers. *Photographic science and photochemistry*. 2002,20:131-148.

[22] Brancaleon L, Moseley H. Laser and non-laser light sources for photodynamic therapy. *Lasers Med. Sci.* 2002;17:173-86.

[23] Wu MC, Hou CY, Jiang CM, Wang YT, Wang CY, Chen HH, Chang HM. A novel approach of LED light radiation improves the antioxidant activity of pea seedlings. *Food Chemistry*. 2007;101:1753-1758.

[24] Tan Y, Xu CS, Xia XS, Yu HP, Bai DQ, He Y, Leung AWN.. Photodynamic action of LED-activated pyropheophorbide-α methyl ester in cisplatin-resistant human ovarian carcinoma cells. *Laser Phys. Lett* 2009;6:321-327.

[25] Yong Tan, Chuan Shan Xu, Xin Shu Xia, He Ping Yu, Ding Qun Bai, Yong He, Jing Xu, Ping Wang, Xin Na Wang, Albert Wing Nang Leung. Preliminary studies on LED-activated pyropheophorbide-α methyl ester killing cisplatin-resistant ovarian carcinoma cells. *Laser Phys.* 2009;19:1045-1049.

[26] Usuda J, Kato H, Okunaka T, Furukawa K, Tsutsui H, Yamada K, Suga Y, Honda H, Nagatsuka Y, Ohira T, Tsuboi M, Hirano T. Photodynamic therapy (PDT) for lung cancers. *J. Thorac. Oncol.* 2006;1:489-93.

[27] Ogata M, Inanami O, Nakajima M, Nakajima T, Hiraoka W, Kuwabara M. Ca2+-dependent and caspase-3-independent apoptosis caused by damage in Golgi apparatus duce to 2,4,5,7-tetrabromorhodamine 123 bromide-induced photodynamic effects. *Photochemistry and Photobiology*, 2003, 78:241-247.

[28] Leung SC, Lo PC, Ng DK, Liu WK, Fung KP, Fong WP. Photodynamic activity of BAM-SiPc, an unsymmetrical bisamino silicon(IV) phthalocyanine, in tumour-bearing nude mice. *Br J. Pharmacol.* 2008;154(1):4-12.

[29] Xu CS, Leung AWN, Liu L, Xia XS. LED-activated pheophorbide a induces cellular destruction of colon cancer cells. *Laser Phys. Lett.* 2010;7:544-548.

[30] Lai JC, Lo PC, Ng DK, Ko WH, Leung SC, Fung KP, Fong WP. BAM-SiPc, a novel agent for photodynamic therapy, induces apoptosis in human hepatocarcinoma HepG2 cells by a direct mitochondrial action. *Cancer Biology and Therapy*. 2006, 5:413-418.

[31] Tournas JA, Lai J, Truitt A, Huang YC, Osann KE, Choi B, Kelly KM. Combined benzoporphyrin derivative monoacid ring photodynamic therapy and pulsed dye laser for port wine stain birthmarks. *Photodiagnosis Photodyn Ther*. 2009;6:195-9.

[32] Tang PM, Bui-Xuan NH, Wong CK, Fong WP, Fung KP. Pheophorbide a-Mediated Photodynamic Therapy Triggers HLA Class I-Restricted Antigen Presentation in Human Hepatocellular Carcinoma. *Transl Oncol*. 2010;3:114-22.

[33] Peloi LS, Soares RR, Biondo CE, Souza VR, Hioka N, Kimura E. Photodynamic effect of light-emitting diode light on cell growth inhibition induced by methylene blue. *J. Biosci.* 2008;33:231-7.

[34] Li W, Yamada I, Masumoto K, Ueda Y, Hashimoto K. Photodynamic therapy with intradermal administration of 5-aminolevulinic acid for port-wine stains. *J. Dermatolog. Treat.* 2010;21:232-9.

[35] Zeng XB, Leung AWN, Xia XS, Yu HP, Bai DQ, Xiang JY, Jiang Y, Xu CS. Effect of blue light radiation on curcumin-induced cell death of breast Cancer cells. *Laser Phys.* 2010; 20:1500-1503.

[36] Bai DQ, Yow CMN, Tan Y, Chu ESM, Xu CS. Photodynamic action of LED-activated nanoscale photosensitizer in nasopharyngeal carcinoma cells. *Laser Phys.* 2010;20: 544-550.
[37] Higuchi A, Yamada H, Yamada E, Jo N, Matsumura M. Hypericin inhibits pathological retinal neovascularization in a mouse model of oxygen-induced retinopathy. *Molecular Vision.* 2008,14:249-254.
[38] Xie TT, Sun LR, Lou HX, Ji M. Research advances of antibacterial action of Hypericum perforatum plants. *Journal of Chinese Medicinal Materials.* 2010;33:146-149.
[39] Xu CS, Leung AWN. Light-activated hypericin induces cellular destruction of nasopharyngeal carcinoma cells. *Laser Phys. Lett.* 2010:7:68-72.
[40] Zhao J, Meng W, Miao P, Yu Z, Li G. Photodynamic effect of hypericin on the conformation and catalytic activity of hemoglobin. *Int. J. Mol. Sci.* 2008;9: 145.
[41] Shen L, Ji H, Zhang H. Anion of hypericin is crucial to understanding the photosensitive features of the pigment. *Bioorganic & Medicinal Chemistry Letters.* 2006;16:1414-1417.
[42] https://bluepoppy.com/cfwebstorefb/index.cfm?fuseaction=feature.display&feature_id=225
[43] Süntar IP, Akkol EK, Yilmazer D, Baykal T, Kirmizibekmez H, Alper M, Yeşilada E. Investigations on the in vivo wound healing potential of Hypericum perforatum L. *Journal of Ethnopharmacology.* 2010;127:468-477.
[44] Karioti A, Bilia AR. Hypericins as potential leads for new therapeutics. *Int. J. Mol. Sci.* 2010;11:562-94.
[45] Bhuvaneswari R, Gan YY, Yee KK, Soo KC, Olivo M. Effect of hypericin-mediated photodynamic therapy on the expression of vascular endothelial growth factor in human nasopharyngeal carcinoma. *International Journal of Molecular Medicine.* 2007, 20: 421-428.
[46] Giese. A. C. Hypericism. Photochem. Photobiol. Rev. 1980; 5:229-255
[47] Chan PS, Koon HK, Wu ZG, Wong RN, Lung ML, Chang CK, Mak NK. Role of p38 MAPKs in hypericin photodynamic therapy-induced apoptosis of nasopharyngeal carcinoma cells. *Photochem Photobiol.* 2009;85:1207-17.
[48] Buytaert E, Matroule JY, Durinck S, Close P, Kocanova S, Vandenheede JR, de Witte PA, Piette J, Agostinis P. Molecular effectors and modulators of hypericin-mediated cell death in bladder cancer cells. *Oncogene.* 2008;27:1916-29.
[49] Miccoli L, Beurdeley-Thomas A, De Pinieux G, Sureau F, Oudard S, Dutrillaux B, Poupon MF. light-induced photoactivation of hypericin affects the energy metabolism of human glioma cells by inhibiting hexokinase bound to mitochondria. *Cancer Res.* 1998,58:5777-5786.
[50] Cole CD, Liu JK, Sheng X, Chin SS, Schmidt MH, Weiss MH, Couldwell WT. Hypericin-mediated photodynamic therapy of pituitary tumors: preclinical study in a GH4C1 rat tumor model. *J. Neurooncol.* 2008;87:255-61.
[51] Kamuhabwa AR, Agostinis PM, D'Hallewin MA, Baert L, de Witte PA. Cellular photodestruction induced by hypericin in AY-27 rat bladder carcinoma cells. *Photochem. Photobiol.* 2001;74:126-32.
[52] Agostinis P, Vantieghem A, Merlevede E, de Witte PA. Hypericin in cancer treatment: more light on the way. *Int. J. Biochem. Cell Biol.* 2002;34:221-241.

[53] Vantieghem A, Xu Y, Assefa Z, Piette J, Vandenheede JR, Merlevede W, De Witte PA, Agostinis P. Phosphorylation of Bcl-2 in G2/M phase-arrested cells following photodynamic therapy with hypericin involves a CDK1-mediated signal and delays the onset of apoptosis. *J. Biol. Chem.* 2002;277:37718-31.

[54] Jäckel MC, Dorudian MA, Marx D, Brinck U, Schauer A, Steiner W. Spontaneous apoptosis in laryngeal squamous cell carcinoma is independent of bcl-2 and bax protein expressio.n. *Cancer* 1999; 85: 591-599.

[55] LaMuraglia GM, Schiereck J, Heckenkamp J, Nigri G, Waterman P, Leszczynski D,Kossodo S. Photodynamic therapy induces apoptosis in intimal hyperplastic arteries. *Am. J. Pathol.* 2000; 157: 867-875.

[56] Ali SM, Chee SK, Yuen GY, Olivo M. Hypericin and hypocrellin induced apoptosis in human mucosal carcinoma cells. *J. Photochem. Photobiol. B.* 2001;65:59.

[57] Chalah A, Khosravi FR. The mitochondrial death pathway. *Adv. Exp. Med. Biol.* 2008;615:25-45.

[58] Smith DJ, Ng H, Kluck RM, Nagley P. The mitochondrial gateway to cell death. *IUBMB Life*. 2008;60:383-389.

[59] Shen Z, Shen J, Li Q, Chen C, Chen J, Zeng Y. Morphological and functional changes of mitochondria in apoptotic esophageal carcinoma cells induced by arsenic trioxide. *World Journal of Gastroenterology*. 2002;8:31-35.

[60] Lee HC, Yin PH, Lu CY, Chi CW, Wei YH. Increase of mitochondria and mitochondrial DNA in response to oxidative stress in human cells. *Biochem J.* 2000;348:425-432.

[61] Chaloupka R, Petit PX, Israël N, Sureau F. Over-expression of Bcl-2 does not protect cells from hypericin photo-induced mitochondrial membrane depolarization, but delays subsequent events in the apoptotic pathway. *FEBS Lett.* 1999;462, 295.

[62] Theodossiou TA, Noronha-Dutra A, Hothersall JS. Mitochondria are a primary target of hypericin phototoxicity: synergy of intracellular calcium mobilisation in cell killing. *Int. J. Biochem. Cell Biol.* 2006;38:1946-56.

[63] Ghavami S, Asoodeh A, Klonisch T, Halayko AJ, Kadkhoda K, Kroczak TJ, Gibson SB, Booy EP, Naderi-Manesh H, Los M. Brevinin-2R semi-selectively kills cancer cells by distinct mechanism, which involves the lysosomal-mitochondrial death pathway. *J. Cell. Mol. Med.* 2008,12:1005-1022.

[64] Kim R. Recent advances in understanding the cell death pathways activated by anticancer therapy. *Cancer*. 2005;103:1551-60.

[65] Price M, Terlecky SR, Kessel D. A role for hydrogen peroxide in the pro-apoptotic effects of photodynamic therapy. *Photochem. Photobiol.* 2009;85:1491-6.

[66] Ali SM, Chee SK, Yuen GY, Olivo M. Hypericin induced death receptor-mediated apoptosis in photoactivated tumor cells. *Int. J. Mol. Med.* 2002;9:601-16.

[67] Smits P, Bolton AD, Funari V, Hong M, Boyden ED, Lu L, Manning DK, Dwyer ND, Moran JL, Prysak M, Merriman B, Nelson SF, Bonafé L, Superti-Furga A, Ikegawa S, Krakow D, Cohn DH, Kirchhausen T, Warman ML, Beier DR. Lethal skeletal dysplasia in mice and humans lacking the golgin GMAP-210. *N. Engl. J. Med.* 2010;362:206-16.

[68] Nakagomi S, Barsoum MJ, Bossy-Wetzel E, Sütterlin C, Malhotra V, Lipton SA. A Golgi fragmentation pathway in neurodegeneration. *Neurobiol Dis.* 2008,29:221-231.

[69] Ran R, Pan R, Lu A, Xu H, Davis RR, Sharp FR. A novel 165-kDa Golgi protein induced by brain ischemia and phosphorylated by Akt protects against apoptosis. *Mol. Cell Neurosci.* 2007,36:392-407.

[70] Chen F, He Y. Caspase-2 mediated apoptotic and necrotic murine macrophage cell death induced by rough Brucella abortus. *Plos one.* 2009,4:e6830-.

[71] Bouchier-Hayes L, Oberst A, McStay GP, Connell S, Tait SW, Dillon CP, Flanagan JM, Beere HM, Green DR.Characterization of cytoplasmic caspase-2 activation by induced proximity. *Molecular Cell.* 2009,35:830-840.

[72] Ferri KF, Kroemer G. Organelle-specific initiation of cell death pathways. *Nat. Cell Biol.* 2001;3:E255-63.

[73] Matroule JY, Carthy CM, Granville DJ, Jolois O, Hunt DW, Piette J. Mechanism of colon cancer cell apoptosis mediated by pyropheophorbide-a methylester photosensitization. *Oncogene.* 2001 5;20:4070-84.

[74] Davids LM, Kleemann B, Cooper S, Kidson SH. Melanomas display increased cytoprotection to hypericin-mediated cytotoxicity through the induction of autophagy. *Cell Biology International*, 2009;33:1065-1072.

[75] Uzdensky AB, Ma LW, Iani V, Hjortland GO, Steen HB, Moan J. Intracellular localisation of hypericin in human glioblastoma and carcinoma cell lines. *Lasers Med. Sci.* 2001;16:276-83.

Reviewer: Tinghe Yu, Email: yutinghe@hotmail.com

Chapter 5

ROLE OF THE TRANS-GOLGI NETWORK (TGN) IN THE SORTING OF NONENZYMIC LYSOSOMAL PROTEINS

*Maryssa Canuel, Libin Yuan and Carlos R. Morales**
Department of Anatomy and Cell Biology, McGill University, Montreal, Canada

ABSTRACT

In eukaryotes the delivery of newly synthesized proteins to the extracellular space, the plasma membrane and the endosome/lysosomal system is dependent on a series of functionally distinct compartments, including the endoplasmic reticulum, the Golgi apparatus and carrier vesicles. This system plays a role in the post-translational modification, sorting and distribution of proteins to their final destination. Most cargo is sorted within, and exits from, the *trans*-Golgi network (TGN). Proteins delivered to the endosomal/lysosomal system include a large and diverse class of hydrolytic enzymes and nonenzymic activator proteins. They are directed away from the cell surface by their binding to mannose-6-phosphate receptors (MPR). Surprisingly, in I-cell disease, in which the MPR pathway is disrupted, the nonenzymic sphingolipid activator proteins (SAPs), prosaposin and GM_2AP, continue to traffic to the lysosomes. This observation led us to the discovery of a new lysosomal sorting receptor, sortilin. Both prosaposin and GM_2AP are secreted or targeted to the lysosomes through an interaction of specific domains with sortilin. In the case of prosaposin, deletion of the C-terminus did not interfere with its secretion, but abolished its transport to the lysosomes. Our investigations also showed that while the lysosomal isomer of prosaposin (65kDa) is Endo H-sensitive, the secretory form (70kDa) is Endo H-resistant. Since the processing pathway within the Golgi apparatus is highly ordered, this Endo-H analysis permitted us to distinguish a sorting sub-compartment where a significant fraction of prosaposin exits to the endosomal/lysosomal system prior to achieving full glycosylation and Endo-H resistance. Mutational analysis revealed that the first half of the prosaposin C-terminus (aa524-540)–contains a saposin-like motif required for its binding to sortilin and its transport to the lysosomes. Additionally, a chimeric construct consisting of albumin and a

---

* Department of Anatomy and Cell Biology, McGill University, 3640 University Street, Montreal, Quebec, Canada, P: 514 398 6398, F: 514 398 5047, E-mail: carlos.morales@mcgill.ca

distal segment of prosaposin, which included its C-terminus, resulted in the routing of albumin to the lysosomes. Based on previous observations showing that the lysosomal trafficking of prosaposin and chimeric albumin required sphingomyelin, we tested the hypothesis that these proteins, as well as sortilin, are associated with detergent-resistant membranes (DRMs). Our results demonstrated that indeed sortilin, prosaposin and chimeric albumin are found in DRMs, and that the sorting of prosaposin to DRMs depends upon the interaction of its C-terminus with sortilin. In conclusion, we have identified a specific segment in the C-terminus of prosaposin, as well as amino acid residues that are critical to the binding of prosaposin to sortilin and its subsequent lysosomal trafficking. The identified sequence may permit the development of new therapeutic approaches for the targeting of proteins with anti-pathogenic properties to penetrate the cell via the endocytic pathway.

I. EVOLUTIONARY SORTING SYSTEMS IN THE TGN

The need to sort proteins between compartments, to the plasma membrane and the extracellular space arose with the development of eukaryotes approximately 2.7 billion years ago [1]. This additional complexity, not present in prokaryotes, is a consequence of the presence of membrane-bound organelles. To cope with the resultant intricacies, a series of functionally distinct compartments, including the endoplasmic reticulum, the Golgi apparatus and transport vesicles evolved to sort proteins destined for secretion or intracellular transport, including to the lysosomes [2, 3].

In mammalians, the mannose 6-phosphate receptors (MPRs) are the canonical sorting receptors responsible for the transport of newly synthesized soluble hydrolases destined for the lysosomes [4]. Two forms of the MPR exist, the 46 kDa cation-dependent (CD) MPR and the 300 kDa cation-independent (CI) MPR [5, 6]. Although there is only modest homology between these two MPRs (~20%), it is accepted that both MPR genes arose from a common ancestor. It has been suggested that the more complex CI-MPR originated from multiple duplications of a single common ancestral gene [7-9]. The MPRs have been described in chickens, amphibians, and reptiles. Although not well characterized, a putative MPR has also been reported in invertebrates [10-12]. However, at present, given the lack of information, it remains difficult to confirm the point of origin in which the ancestral MPR gene appeared during evolution [13]. Nonetheless, recent findings have demonstrated that the MPR is present and fully functional in ancient teleosts which appeared during the Triassic period over 200 million years ago [13, 14]. As eukaryotes and the need to sort proteins arose at a much earlier point in time, the MPR would either need to have arisen before the Triassic period, or another more ancient sorting receptor must exist.

The quest for sorting receptors in yeast and plants resulted in the discovery of the Vps10p and the BP80 receptors [15, 16]. The Vps10 receptor family is a novel class of heterogeneous type-I trans-membrane receptors which includes sortilin (100 kDa), Vps10p (160 kDa), SorLA (250 kDa), and SorCS1-3 (130kDa) [17]. Vps10 family members all have names beginning with the pre-fix "sor" which is an abbreviation of "sorting receptor related" and highlights the functional role of these proteins [18]. These receptors have a diverse pattern of expression in many tissues and are responsible for the targeting of a variety of different ligands [19-24]. The explanation for the existence of multiple lysosomal sorting pathways may be evolutionary. As Vps10 domain containing proteins are found in a variety of simple organisms such as Dictyostelium, Neurospora, and Metarhizium, it is probable that sorting pathways involving these receptors arose before that of the MPR [25].

Sortilin, a well-studied member of the Vps10 family, is a 100 kDa sorting receptor that is highly expressed in the brain, testis and skeletal muscle [24]. While sortilin was known to bind and internalize proteins at the cell surface, our laboratory was the first to identify intracellular ligands, namely, prosaposin and the $G_{M2}AP$, that require sortilin for their intracellular transport to the lysosomes [21, 26].

This chapter will focus on the pathways and processes involved in regulating the sorting of prosaposin and $G_{M2}AP$ from the TGN to the lysosomes. Specifically, we will examine sortilin, the sorting receptor responsible for the lysosomal transport of prosaposin and $G_{M2}AP$, as well as the sequences that mediate their interaction, post-translational modifications, and localization to detergent-resistant membranes (DRMs). A better understanding of the mechanisms involved in the targeting of these proteins to the lysosomes may permit the development of new therapeutic approaches to target proteins with anti-pathogenic properties to the endocytic compartment.

II. THE FUNCTION OF LYSOSOMAL PROTEINS

Overview

It is well known that the plasma membrane of mammalian cells continuously flows and recycles between the cell surface and organelles through a vesicular system [27]. During endocytosis, small portions of the plasma membrane bud into coated and none coated pits, which are internalized as vesicles and transported from the cell surface to the endosomal/lysosomal system [28]. Lysosomes are the major degradation compartments in eukaryotic cells and contain a large variety of acid hydrolases to digest proteins, lipids, nucleic acids, and carbohydrate molecules [29]. In the lysosomes, part of the plasma membrane is incorporated in the lysosomal membrane, and is protected from the action of hydrolases due to the presence of a thick sugar coat on the luminal surface of the membrane surrounding these organelles [30, 31]. This coat corresponds to the carbohydrate portion of N-glycosylated lysosomal integral membrane proteins (LIMPs) and lysosomal associated membrane proteins (LAMPs) [30, 31]. Eventually, part of the internalized membrane recycles back to the cell surface carrying certain receptors that escape from degradation [27]. This recycling process enhances the ability of the cell surface to internalize ligands and other materials [27]. However, a subset of plasma membrane is destined for degradation by lysosomal hydrolases [32]. As a consequence, some regions of the lysosomal membrane invaginate into the lumen of the endosomes and multivesicular bodies to form intra-endosomal vesicles [28]. The convex curvature of these small vesicles (40-100 nm in diameter) favors the spreading of head groups of the glycosphingolipids on the outer leaflet of intra-endosomal/lysosomal vesicles membrane and makes them more easily accessible for the lysosomal hydrolases [33]. Due to an endosomal pH decrease, the composition of the internal membrane changes and most of the membrane-stabilizing cholesterol is removed [34, 35]. Bis (monoacylglycero) phosphate (BMP), a negatively charged phosphate lipid which favors a strong curvature, is also increased to facilitate the membrane degradation [34, 35]. Through this digestive process, glycosphingolipids are cleaved into small molecules by glycosidases, with the stepwise release of sphingosine, fatty acid, monosaccharides, glycerol, sulfate and sialic acids [36].

Lysosomal glycosidases are water-soluble enzymes with the ability to bind to negatively-charged membrane or water-soluble substrates [36]. They are also able to digest membrane-bound substrates with long sugar chains that reach the aqueous phase [37]. However, glycosphingolipids with short oligosaccharide side chains (four or less sugar residues) cannot access the active sites of the glycosidases [28]. Their degradation requires sphingolipid activator proteins (SAPs) and negatively charged lysosomal lipids, which disturb the membrane structure, extract the sphingolipids from the membrane and expose them to the glycosidases [38, 39].

Lysosomal Functions of Sphingolipid Activator Proteins (SAPs)

Five SAPs have been described: the $G_{M2}AP$ and four saposins [40, 41]. The $G_{M2}AP$ is a 17.6kD glycoprotein in its deglycosylated form and is encoded by a gene on chromosome 5 [42, 43]. It acts as an essential activator for the degradation of G_{M2} by N-acetyl-β-hexosaminidase A [44]. During the process of hydrolysis, $G_{M2}AP$ helps the enzyme to cleave a GalNAc and a NeuAc residue from G_{M2} and convert this glycosphingolipid to G_{M3} [45].

It has been proposed that N-acetyl-β-hexosaminidase A and sialidase cannot cleave G_{M2} gangliosides on membrane surfaces because the oligosaccharide of the glycosphingolipids does not extend far enough into the aqueous phase [39]. The $G_{M2}AP$ appears to contain a hydrophobic cavity that harbors the ceramide moiety of G_{M2} ganglioside [39]. According to a proposed model, the $G_{M2}AP$ inserts into the bilayer of intra-lysosomal vesicles, lifts the G_{M2} ganglioside out of the membrane, and presents it to the active site of β-hexosaminidase A [39]. In fact, the $G_{M2}AP$ shows strong ability to bind to acidic glycosphingolipids, including gangliosides and sulfated glycosphingolipids [45]. The negatively charged sugar residues or sulfate groups in gangliosides appear to be important for the binding to $G_{M2}AP$ [45].

The other four activator proteins are the saposins, which are individually termed saposin A, B, C and D. Saposins are acidic, heat-stable glycoproteins with a size of 13 to 15 kDa [46]. They are derived from a common precursor, prosaposin, encoded by a gene on chromosome 10. In the central segment of the prosaposin backbone, the saposins are arranged in tandem and classified as B-type domains [47, 48]. All four saposins show homology to each other [49], however, they have different substrate specificities [48]. Saposin A is required for the degradation of galactosylceramide by galactosylceramidase [50]. Saposin B shows a broader specificity, but mainly for sulfatide by arylsulfatase A *in vivo* [36]. Saposin C is required for the lysosomal degradation of glucosylceramide by glucosylceramidase, and saposin D stimulates lysosomal ceramide degradation by acid ceramidase [51, 52].

Inherited deficiencies of SAPs lead to the accumulation of undegradable membranes within the lysosomal compartment and to the development of lysosomal storage disorders (LSD), such as Krabbe disease (saposin A deficiency), metachromatic leukodystrophy (saposin B deficiency), and Gaucher's disease (saposin C deficiency) [53-56]. Deficiency of the $G_{M2}AP$ blocks the degradation of G_{M2} ganglioside and results in an AB variant G_{M2}-gangliosidosis, characterized by the accumulation of G_{M2} and GA2 gangliosides in the lysosomes of neuronal cells [57]. Therefore, a complex but necessary mechanism that regulates the selective degradation of glycosphingolipids of short oligosaccharides side chains is critical for membrane catabolism. Thus, SAPs are essential factors in membrane turnover that function by facilitating the catalysis of certain sphingolipids by their specific hydrolases.

III. SORTING OF LYSOSOMAL PROTEINS

Lysosomal transport occurs by two separate mechanisms depending on the physical nature of a protein. Integral membrane proteins interact directly with adaptor proteins to recruit clathrin and traffic to the lysosomes whereas soluble proteins must interact with trans-membrane lysosomal sorting receptors which in turn mediate lysosomal transport. The MPRs and sortilin are the major sorting receptors involved in Golgi to lysosomal trafficking.

The Mannose-6-Phosphate Receptors (MPRs)

The MPRs are responsible for routing most soluble proteins to the lysosomes [4]. MPR ligands comprise most soluble hydrolases, including cathepsins and various sulfatases, IGF-II, pro-renin, renin, granzyme B, proliferin, retinoic acid, uPAR, and plasminogen [4, 58-64]. While both the CD- and CI-MPR are implicated in lysosomal sorting, the CI-MPR is the dominant lysosomal targeting receptor for soluble hydrolases [65, 66]. The recognition of proteins by the MPR occurs in the TGN and is dependent upon the presence of a mannose 6-phosphae (M6P) residue on the newly synthesized enzyme [67-69]. The addition of the M6P tag occurs in a two-step process initiated by the addition of N-acetylglucosamine-1-phosphate to mannose residues by an N-acetylglucosamine-1-phosphotransferase [70, 71]. Trimming of N-acetylglucosamine residues by N-acetylglucosamine-1-phosphodiester α-N-acetylglucosaminidase results in the exposure of the M6P tag [72]. The M6P tag and the soluble protein are then ready for recognition and trafficking by the MPR. The receptor-ligand complex remains intact and is transported to the endosomal compartment wherein the acidic environment causes dissociation of the ligand from the MPR [73].

Sortilin

While it is clear that the MPR is the canonical lysosomal sorting receptor, it is equally evident that an alternative pathway also exists for certain lysosomal proteins. Multiple lines of evidence support this theory, including that of fibroblasts from patients with mucolipidosis type II/I-Cell disease (ICD). Patients with ICD lack a functional MPR sorting pathway as a result of a mutation of the UDP-N-acetylglucoasmine-1-phosphotransferase that is responsible for the addition of M6P residues. Despite this, certain soluble lysosomal proteins, such as the SAPs and cathepsin D, are transported to the lysosomes [71, 74]. Moreover, metabolically labeled rat hepatocytes revealed that despite the presence of an M6P tag, cathepsin H was secreted into the medium with no correlate targeting to the lysosomes [75]. Similarly, competition assays demonstrated that the addition of exogenous M6P did not alter the subcellular localization of cathepsin K. Cathepsin L may also traffic to the lysosomes independently of the MPR. While an M6P is present on cathepsin L, this soluble lysosomal hydrolase was found to have low affinity to the MPR and depend on a proteinaceous signal for its sorting [76-78]. These results provide further evidence of an MPR-independent pathway [79]. Recently, the sortilin sorting receptor has been implicated in MPR-independent trafficking [21, 80, 81].

Sortilin, also known as the neurotensin receptor-3, is a member of the Vps10 family of sorting receptors. Sortilin is a 100 kDa protein composed of a luminal/extracellular region containing a cysteine-rich domain Vps10 domain, a trans-membrane domain and a short cytosolic domain that contains a signal for rapid internalization and trafficking [17, 26, 82]. The luminal domain of sortilin contains a furin cleavage site (RWRR) that when recognized in the Golgi apparatus results in the conversion of sortilin to its mature form. The cleaved propeptide segment interacts with the processed receptor and competes with RAP and neurotensin for binding to sortilin. The interaction of the propeptide with sortilin has been hypothesized to favour the proper folding of sortilin, as well as functioning as a molecular safeguard against the formation of death-signalling protein complexes involving sortilin [17, 23]. Within the cell, sortilin is localized primary to the TGN, however approximately 10% of the receptor may be found at the cell surface. Thus, sortilin is implicated in the sorting of ligands to various intracellular locations [24]. Proteins sorted by sortilin include: neurotensin, nerve growth factor precursor (pro-NGF), lipoprotein lipase (LpL), SAPs, acid sphingomyelinase, and cathepsins D and H [21-24, 80, 81]. The cytoplasmic domain of sortilin is essential in mediating the localization of this receptor as well as in its ability to appropriately sort proteins. The cytosolic domain of sortilin is highly homologous to that of the CI-M6PR. This segment contains both YXXϕ (Tyr- any two amino acids- bulky hydrophobic amino acid) and DXXLL (dileucine) sorting motifs that are required for endocytosis and Golgi to endosomal sorting. These sorting motifs function by recruiting and interacting with the necessary adaptor proteins and complexes [82-84].

Adaptor Proteins – Overview

Sorting of proteins to appropriate cellular locations depends upon adaptor proteins. Adaptor proteins provide a link between sorting receptors, like the MPRs and sortilin, and cytoplasmic components implicated in cellular targeting such as clathrin, Arf1, γ-synergin, epsinR, and others [85]. Transport between the TGN and the endosomal compartment may be divided into two distinct processes, forward transport, known as anterograde trafficking, and reverse or retrograde transport from the endosomes to the TGN.

Adaptor Proteins - Anterograde Transport

Based upon their structure, two general classes of adaptor proteins have been demonstrated to be essential in the trafficking of proteins between the TGN and the endosomes, namely the multimeric adaptor proteins and the monomeric adaptor proteins also termed Golgi-localized, γ-ear containing, Arf binding proteins (GGAs) [86, 87]. The multimeric adaptor protein (AP) family includes AP-1, AP-2, AP-3, and AP-4 [88].

To be sorted from the Golgi apparatus, the MPRs and sortilin must bind to GGAs [89]. The GGAs are a family of three similar proteins (GGAs 1, 2 and 3) that were first identified in a database search for AP subunit homologues [87, 89, 90]. GGAs are composed of four domains: an N-terminal VHS domain, a GAT domain, a hinge domain, and an EAR domain [91]. These proteins range in size from 60-80kDa.

The VHS domain is implicated in mediating the interaction of the GGA adaptors with the cytoplasmic tails of sortilin and the MPR through a DXXLL motif found in these sorting receptors [82, 84, 92]. The GAT domain, the most highly conserved domain, interacts with GTP-bound Arfs to mediate the recruitment of GGAs to the Golgi membrane [87, 93]. The GGA hinge domain interacts with clathrin, whereas the EAR domain binds vesicular coat proteins such as clathrin, γ-synergin, p56, and rabaptin-5 [87, 94].

GGAs are essential for the transport of sortilin and the MPRs from the TGN to the endosomes [21, 82].

A dominant-negative GGA-3 construct lacking the hinge and ear domains abrogated transport of sortilin from the TGN and in fact resulted in its accumulation in the perinuclear region [21]. Depletion of the guanine nucleotide exchange factor (GEF), GBF1, has been suggested to inhibit the recruitment of GGAs to the *cis*- and *mid*-Golgi compartments and the normal processing of prosaposin [95]. However, the prevailing view is that GBF1 is in fact involved in ER-to-Golgi trafficking via the regulation of COPI recruitment in the early secretory pathway [96-98]. Thus, it is likely that another GEF may be involved in the recruitment of GGAs to the TGN.

The multimeric adaptor proteins, and in particular AP-1, are also involved in the anterograde transport of sortilin and the MPR. AP-1, like the other multimeric adaptor proteins is composed of four subunits: two large subunits (γ, α, δ, or ε, and β), a medium subunit (μ), and a small subunit (σ). The four subunits organize into globular structures, wherein the core is composed of the N-termini of the large subunits and the medium and small subunits. The C-terminal domains of the two large subunits form the "ears" of the adaptor protein [99].

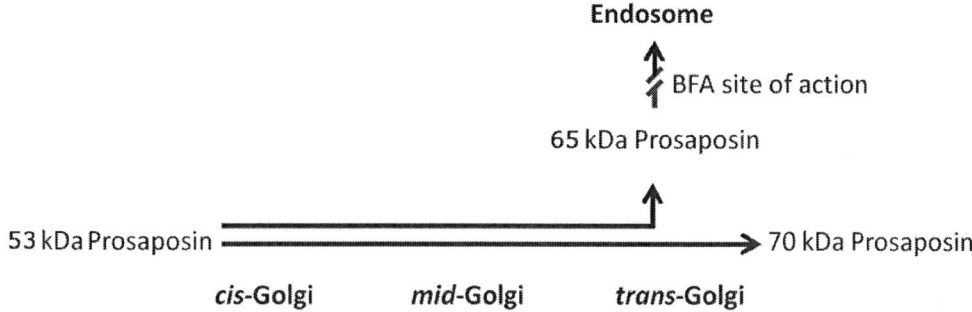

Figure 1. Functions of the TGN. 1) Terminal glycosylation of (70 kDa) prosaposin and sorting to the secretory system; 2) Sorting of lysosomal (65 kDa) prosaposin by sortilin; 3) Recruitment of accessory proteins such as the GGAs and APs; 4) Site of action of GEF proteins; 5) Recruitment of clathrin; 6) Formation of sortilin/prosaposin complexes destined to the endosomes (forward transport); 7) Site of entry of sorting receptors via retromer (reverse transport).

AP-1 exists in two isoforms: AP1-A and AP-B. AP-1A plays an important in sorting events that occur at the TGN, whereas AP-1B is implicated in sorting integral membrane proteins from the TGN or endosomes to the basolateral domains [100].

The μ subunit of AP-1 contains a YXXϕ motif that has been demonstrated to mediate an interaction with cytosolic tail of the MPR [101-103]. Given the high degree of homology between the cytosolic domains of sortilin and the MPR, investigations were conducted to determine if AP-1 is implicated in sortilin sorting. Residues 789-799 of sortilin were found to mediate an interaction with the μ subunit of AP-1 and siRNA inhibition of AP-1 resulted in the abrogation of the lysosomal transport of sortilin and its accumulation in the TGN. Thus, AP-1 is essential in the anterograde trafficking of sortilin [103]. Despite this, in the past there has been controversy over the exact role of GGAs and APs in mediating TGN to endosomal sorting [21, 104]. Much of the controversy arose from findings that fibroblasts deficient in the μ subunit of AP-1 have normal Golgi to endosomal sorting [105]. However, it now appears likely that GGAs and AP-1 function together to recruit cargo and that AP-1 then mediates transport. Evidence for this model includes findings that AP-1, but not GGAs, are enriched in purified clathrin-coated vesicles, and that the membrane association of AP-1 is stable while that of the GGAs is transient [86, 89]. However, it is now clear that both GGAs and AP-1 are important for the efficient delivery of sorting receptors and their cargo to the lysosomes (Figure 1).

Adaptor Proteins - Retrograde Transport

The recycling of sorting receptors is essential for the normal function of lysosomes [106, 107]. The pentameric retromer complex is an important mediator of recycling for the MPR, Vps10p and sortilin sorting receptors from the endosomes to the TGN [103, 107, 108]. The mammalian retromer complex is composed of Vps26, Vps29, Vps35 and sorting nexins 1 and 2 (SNX1 and SNX2) subunits [106, 109, 110]. The Vps35 and Vps26 subunits are responsible for the interaction of the complex with sorting receptors, whereas other subunits are implicated in interactions with the endosomal membrane [107, 108, 111]. Several conserved sequences have been demonstrated to mediate the interaction of retromer with the MPR and sortilin sorting receptors. Mutation of a WLM motif in the cytosolic tail of the MPR attenuated its ability to interact with retromer, as well as its ability to be recycled to the TGN, leading to its rapid degradation. In the cytosolic tail of sortilin, a similar motif, FLV, was postulated to mediate the interaction of sortilin with retromer [107]. Yeast Two-Hybrid experiments demonstrated that amino acids 789-799, a segment that does not contain the FLV segment, are involved in the interaction of sortilin and the Vps35 subunit of retromer [103]. Instead, amino acids 789-799 contained a YXXΦ sorting motif. Furthermore, mutation of Y14 or L17 inhibited the ability of sortilin to interact with retromer. Knock-downs of the Vps26 subunit resulted in the depletion of sortilin in the Golgi apparatus, indicating that a functional retromer complex is required for the recycling of sortilin from the endosomes to the TGN [103]. However, there are other mediators of receptor recycling. Examples of these include syntaxins 5 and 16. SiRNA inhibition of these proteins resulted in the blockage of MPR recycling and an increase in its degradation. Similarly, PACS-1, TIP-47, and epsinR have been demonstrated to mediate recycling from the endosomal compartment [82, 112, 113].

IV. PROSAPOSIN TRAFFICKING

Overview

Prosaposin is a glycoprotein with different functions and destinations. Prosaposin is synthesized in the ER as a 53 kDa protein and post-translationally modified to a 65 kDa form after the addition of high mannose [114-116]. In the TGN, this protein is further glycosylated to a 70 kDa secretory form that is found in various fluids (Figures 1 and 2A). In the nervous system, secretory prosaposin may function as a neurotrophic factor [117]. Unlike the 70 kDa form, the 65 kDa protein is associated with the membrane of the Golgi apparatus, where it is sorted and targeted to lysosomes (Figures 1 and 2B) [118]. In fact, permeabilization of Golgi-enriched fractions with saponin liberated the 70 kDa form but not the 65 kDa protein, and excess free M6P did not release lysosomal prosaposin from Golgi membranes (Figure 2B) [118]. Furthermore, quantitative electron microscopy demonstrated that the lysosomal content of prosaposin increased significantly after administration of tunicamycin [118]. These results constituted the first clear indication that the trafficking of the 65 kDa form of prosaposin to the lysosomes was independent of the MPR pathway.

One of the main mechanisms for regulating the sorting of secretory proteins within the TGN is their selective aggregation in response to a mild acidic pH and an elevated Ca^{2+} concentration [119].

The 70 kDa form of prosaposin was found to aggregate within the permeabilized Golgi fractions in a pH and Ca^{2+} dependent manner (Figures 2C and D). Enriched Golgi fractions incubated at pH 5.4 and 7.4 with saponin released the 70 kDa form, whereas a pH of 6.4 caused retention of this protein (Figure 1D). Thus, prosaposin is initially glycosylated as a membrane-associated 65 kDa protein which can either be transported to the lysosomes or further processed into a highly glycosylated 70 kDa form. The 70 kDa protein readily aggregates within the TGN and is then secreted, possibly in a regulated manner.

The 65 kDa isomer is transported to the lysosomes and submitted to partial proteolysis resulting in four smaller non-enzymatic saposin molecules which are implicated in the hydrolysis of sphingolipids [120-122]. The Golgi apparatus, however, is not only responsible for accomplishing the molecular sorting of the 70 kDa form, but also for decoding the lysosomal sorting signal from the amino acid backbone of the 65 kDa form of prosaposin.

The identification of the prosaposin region involved in this interaction was found in our laboratory by mutational deletion of each of the four known saposin functional domains and the highly conserved N- and C- termini of prosaposin [49, 116]. The truncated cDNAs were subcloned into pcDNA3.1B expression vectors and transfected into COS-7 cells.

The myc-tagged truncated proteins were detected by immunofluorescence using anti-myc antibody, followed by a secondary FITC conjugated goat anti-mouse antibody [116]. We found that deletion of the C-terminus did not interfere with the routing of prosaposin to the extracellular compartment but abolished its transport to lysosomes.

Deletion of each of the saposin regions and N-terminal domain did not affect the lysosomal or secretory routing of prosaposin [116].

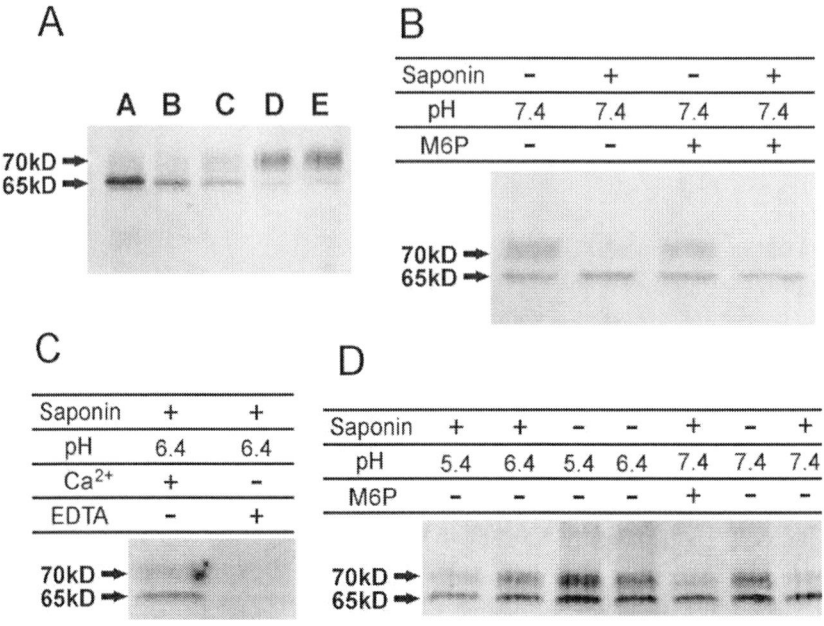

Figure 2. Effect of Golgi permeabilization on prosaposin under different experimental conditions. A) Time course labeling of prosaposin in Sertoli cells. Lanes A, B,C, D and E, represent 30, 60, 120, 180 and 240 min respectively after exposure to ^{35}S-cysteine. The 70 kDa form of prosaposin represents a post-translational modification of the 65 kDa polypeptide. B) Golgi fractions of Sertoli cells permeabilized at pH 7.4 in the presence or absence of M6P and immunoprecipitated with prosaposin antibody. Permeabilization in presence of free M6P did not release the 65 kDa form indicating that it is not associated with the MPRs. C) Permeabilization at pH 6.4 caused the retention of the 70 kDa form. EDTA released the 70 and 65 kDa forms, indicating that the aggregation of the 70 kDa protein at pH 6.4 is selective and enhanced in presence of calcium ions. D) Permeabilization of Golgi fractions at neutral pH (7.4) released the 70 kDa form, but not the 65 kDa form of prosaposin. Permeabilization at pH 6.4 or lower caused the retention of the 70 kDa. These results suggest that the 70 kDa protein aggregates in the TGN. *(Reproduced with permission from Igdoura S.A., Rasky A., Morales C.R. Trafficking of sulfated glycoprotein-1/prosaposin to lysosomes or to the extracellular space in rat Sertoli cells. Cell Tiss. Res. 283: 385-394, 1986)*

To determine whether the C-terminus alone was sufficient for the intracellular targeting of this protein, a chimeric construct encoding the full-length albumin plus the C-terminus of prosaposin (Alb/COOH) was engineered and subcloned into the pcDNA3.1B vector, and transfected into COS-7 cells. A wild type albumin cDNA was also prepared as a control (Alb). After transfection, cells were immunostained with anti-myc antibody, followed by a secondary FITC conjugated goat anti-mouse antibody. The cells were simultaneously stained with the lysosomal marker LysoTracker. The results showed that the C-terminus alone was insufficient to target the chimeric construct to the lysosomes [116].

Based on this result and on the mutational analysis of prosaposin, we hypothesized that perhaps one or more saposin domains are required, along with the C-terminus to direct albumin to the lysosomes. To test this hypothesis, a chimeric protein was engineered by fusing an albumin cDNA with a cDNA sequence encoding the domain D and the C-terminus of prosaposin (Alb/D/COOH). In addition, an albumin fusion protein containing domains C and D plus the C-terminus of prosaposin (Alb/C/D/COOH) was also constructed.

In these cases, the anti-myc antibody yielded a punctuate reaction (Figure 3) which overlaid with LysoTracker staining. Western blotting with anti-myc antibody confirmed the expression of wild type and chimeric albumin proteins. This observation indicated that the C-terminus and at least one saposin domain was required to direct albumin to the lysosomal compartment [116].

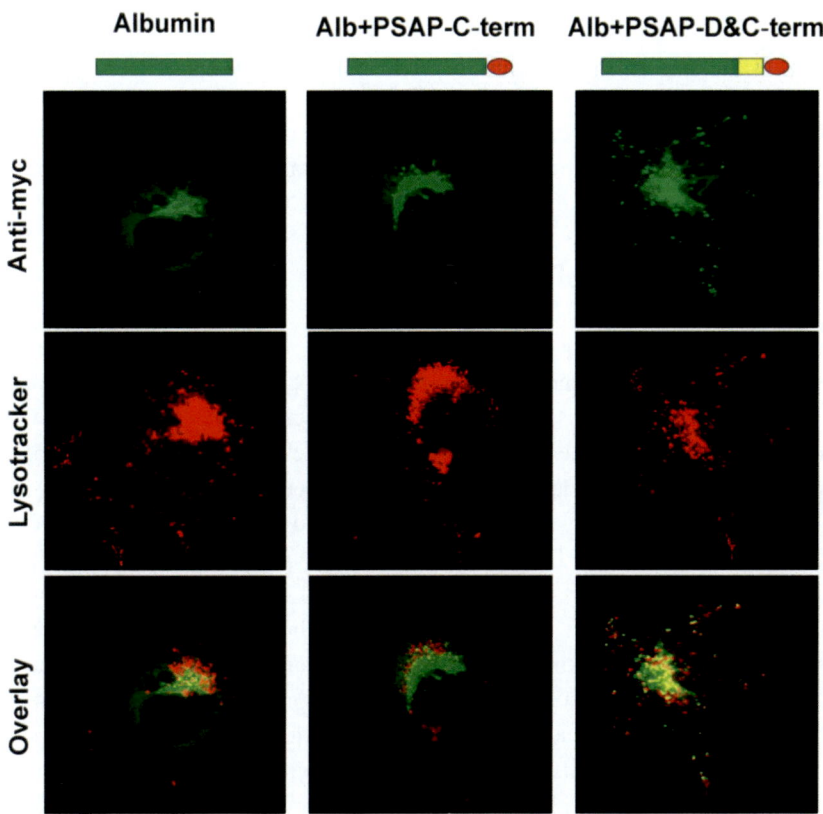

Figure 3. Targeting of albumin/prosaposin chimeric proteins to lysosomes. COS-7 cells were transfected with albumin, albumin-prosaposin-C-terminus (Alb+PSAP-C-term), and albumin-prosaposin-D-domain followed by the C-terminus (Alb+PSAP-D and C-term) constructs linked to myc. The cells were stained with anti-myc (green fluorescence) and counter stained with Lyso-Tracker (red fluorescence). The punctuate reaction obtained with anti-myc in Alb+PSAP-D and C-term overlapped with Lysotracker and demonstrating that the prosaposin D domain followed by the C-terminus is required for the transport of albumin to lysosomes. *(Reproduced with permission from Zhao Q., Morales C.R. Identification of a novel sequence involved in lysosomal sorting of the sphingolipid activator protein prosaposin. J. Biol. Chem. 275: 24829-24839, 2000)*

Strong evidence from other laboratories also supported the hypothesis that prosaposin was targeted to lysosomes in a MPR-independent manner [74, 118, 123]. Evidence for a MPR-independent transport of prosaposin included experiments in cultured cells incubated with fumonisin B1, an inhibitor of sphingolipid synthesis that competes with sphinganine as a substrate of ceramide synthase [124]. This treatment produced a dramatic decrease in the immunogold labeling of lysosomes with anti-prosaposin antibody [124].

To examine if the MPR-mediated pathway was affected by this treatment, cells treated or not with fumonisin B1 were labeled with anti-cathepsin A antibody. The results showed no significant differences in the immunogold labeling of the lysosomal compartment of the treated and untreated cells indicating that the effect of fumonisin B1 on the transport of prosaposin to the lysosomes was specific [124]. The effect of DL-threo-1-phenyl-2-decanoyl-amino-3-morpholino-1-propanol-HCL (PDMP), a compound that selectively inhibits the synthesis of glycosphingolipids, but not of sphingomyelin and/or ceramide, and the effect of tricyclodecan-9-yl xanthate potassium (D609), which specifically blocks the formation of sphingomyelin [125] were also examined. The results showed that only D609 blocked the transport of prosaposin to the lysosomes, suggesting that sphingomyelin was the main sphingolipid implicated in the trafficking of prosaposin to the lysosomes [124]. Taken together, these observations suggested that the 65 kDa lysosomal isomer remained associated to an alternative receptor possibly on discreet membrane microdomains such as lipid rafts.

Sortilin-Mediated Sorting of Prosaposin

The sortilin sorting receptor was demonstrated to be responsible for mediating the transport of prosaposin to the endosomal compartment [21]. This was determined using a dominant-negative competition assay and co-immunoprecipitation (Co-IP). In the dominant-negative experiments, a truncated sortilin construct lacking its cytosolic domain that is required to bind and recruit adaptor proteins was over-expressed in COS-7 cells. This construct abolished the ability of prosaposin to traffic to the lysosomes and resulted in its retention in the Golgi apparatus. Despite this, a majority of prosaposin was secreted into the extracellular medium suggesting that upon saturation of the sortilin receptors with ligand, prosaposin may enter a default secretory pathway. The effect of the truncated sortilin construct was specific, as trafficking of cathepsin B, a ligand of the MPR, was unaffected. To validate these findings, the lysosomal trafficking of prosaposin was examined in cells deficient in sortilin. This was achieved by the knockdown of sortilin using siRNA. As expected, in sortilin-deficient cells prosaposin was not routed to the lysosomal compartment. Rather, pulse-chase analysis revealed that in these cells, prosaposin secretion was augmented. The ability of prosaposin to interact directly with sortilin was assessed using an *in vitro* Co-IP assay. Prosaposin, full-length, and truncated sortilin were translated *in vitro* using a reticulocyte lysate system. Prosaposin was then incubated with either full-length or truncated sortilin and Co-IP performed. The results demonstrated that prosaposin immunoprecipitated both full-length and truncated sortilin. Cathepsin B however did not immunoprecipitate either form of sortilin. Thus, sortilin was demonstrated to interact directly with prosaposin and be required to mediate the transport of this SAP to the lysosomal compartment [21].

Prosaposin Processing in the Golgi Apparatus

Effect of Endo H Treatment
The processing of glycoproteins within the Golgi apparatus is highly ordered and treatment with endoglycosidase H (Endo H) is frequently used to distinguish complex from high mannose and hybrid oligosaccharides linked to glycoproteins [73, 126]. The difference

between the 65 kDa and 70 kDa forms of prosaposin reflects variations in its glycosylation state [127]. Thus, to determine the structures of the N-linked carbohydrates and the potential sorting compartment of prosaposin within the Golgi apparatus, Endo H digestion was performed on immunoprecipitated proteins with an anti-myc antibody from lysates of COS-7 cells transfected with a prosaposin-myc construct. The results revealed that the 65 kDa protein was converted into smaller peptides bands [127].

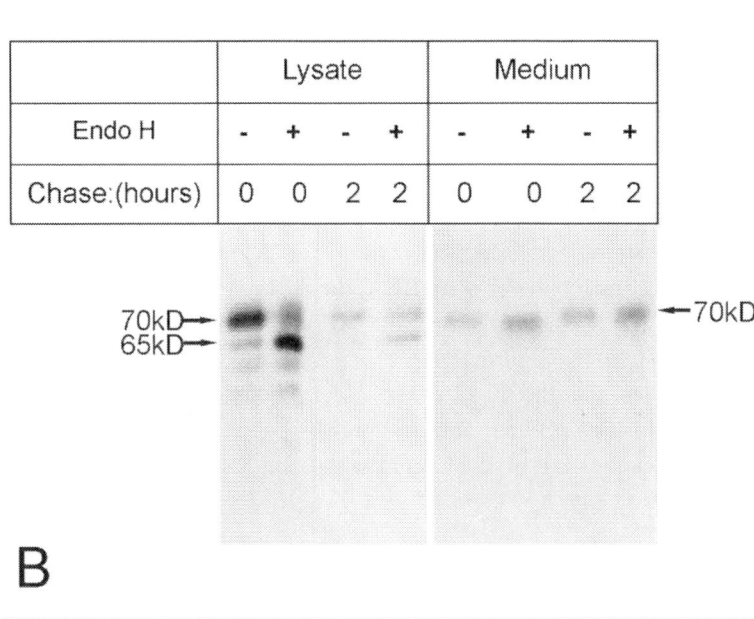

Figure 4. Effects of Endo H and BFA treatments. A) COS-7 cells transfected with prosaposin-myc were labeled with Tran ^{35}S-label and chased for 2 hours. Cell lysates and culture medium were immunoprecipitated with anti-myc antibody and incubated with Endo H. While the 65 kDa protein was converted to smaller peptides bands, the 70 kDa protein was resistant to Endo H treatment. The 70 kDa isomer immunoprecipitated from the medium was also Endo H resistant. B) COS-7 cells transfected with prosaposin-myc were labeled with Tran ^{35}S-label and grown in presence or absence (control) of BFA. BFA caused intracellular retention of the 65 kDA protein. In control cells, the 65 kDa protein decreased intracellularly, while the 70 kDa increased in the medium. *(Reproduced with permission from Zhao Q., Morales C.R. Identification of a novel sequence involved in lysosomal sorting of the sphingolipid activator protein prosaposin. J. Biol. Chem. 275: 24829-24839, 2000)*

On the other hand, the band corresponding to the 70 kDa protein was resistant to Endo H treatment (Figure 4A). Similarly, the 70 kDa secretory protein immunoprecipitated from the medium was Endo H resistant (Figure 4B) [116]. This observation suggested that while the 65 kDa lysosomal form of prosaposin exited from the medial Golgi compartment and/or from a distinct region of the TGN, the 70 kDa protein traversed the Golgi apparatus acquiring terminal glycosylation (Figure 1).

Effect of Brefeldin A (BFA) Treatment

BFA is a fungal metabolite that causes rapid redistribution of the *cis* and *mid* compartments of the Golgi apparatus to the ER and is frequently employed to examine the effect of a disrupted Golgi on the trafficking and sorting of glycoproteins [128, 129].

COS-7 cells transfected with a wild-type prosaposin (PSAP-WT) were incubated with BFA 1 hour prior to trans^{35}S-labeling. After labeling for 30 min, the cells were chased for an additional 30 min, 2h, and 4 h. In presence of BFA, the amount of intracellular prosaposin (65 kDa) in cell lysates remained unchanged after chasing for 4h, while in cells not treated with BFA, the amount of protein decreased after chasing for 2h (Figure 4B).

In the medium of cells incubated with BFA, there was a negligible amount of the 70 kDa protein. In contrast, in cells incubated without BFA, the amount of protein in the medium increased after 30 min and 2h (Figure 4B).

These results suggest that the 65 kDa lysosomal precursor contains high mannose/hybrid sugar residues added in the proximal and mid regions of the Golgi apparatus and that the 70 kDa secretory form achieves full glycosylation with complex sugar residues and exit from the most distal region of the Golgi apparatus (Figure 1) [116].

Effects of Sequential Truncations of the Prosaposin C-Terminus

To identify the specific region within the prosaposin C-terminus that binds sortilin, we generated several truncated prosaposin constructs (Figure 5A) [130]. Before engineering these constructs, we analyzed the predicted secondary structure of the C-terminus using EMBOSS Garnier [131]. The EMBOSS output file predicted the existence of two α-helices within the C-terminus. The first helix was localized to the linker region between aa518-523, and the second between aa540-550. The PredictProtein software, suggested that two pairs of cysteine residues (C528-C536 and C545-C551) may form two disulfide bonds that stabilize the tertiary structure of the C-terminus [132].

Thus, to generate the first construct, termed P-75, and to avoid disruption of the predicted helices and disulfide bonds, we deleted the C-terminal region located immediately after C551. In the second construct, termed P-50, we deleted the region spanning between aa541-557 and eliminated the second helix (E540 to H550). In the third construct, termed P-25, the deletion spanned between aa531-557, resulting in the elimination of two cysteine residues. We also generated a construct, termed P-0, which lacked the entire A-type domain (aa524-557). The final construct, P-ΔC, was a truncated prosaposin lacking the entire C-terminus and the linker that connects this region to saposin D (aa518-557). Subsequently, COS-7 cells were co-transfected with sortilin and each of the prosaposin constructs described above. The cells were homogenized in lysis buffer (pH 6.0) and subjected to immunoprecipitation. The complexes were pulled down with anti-sortilin antibody and resolved on a 10% acrylamide

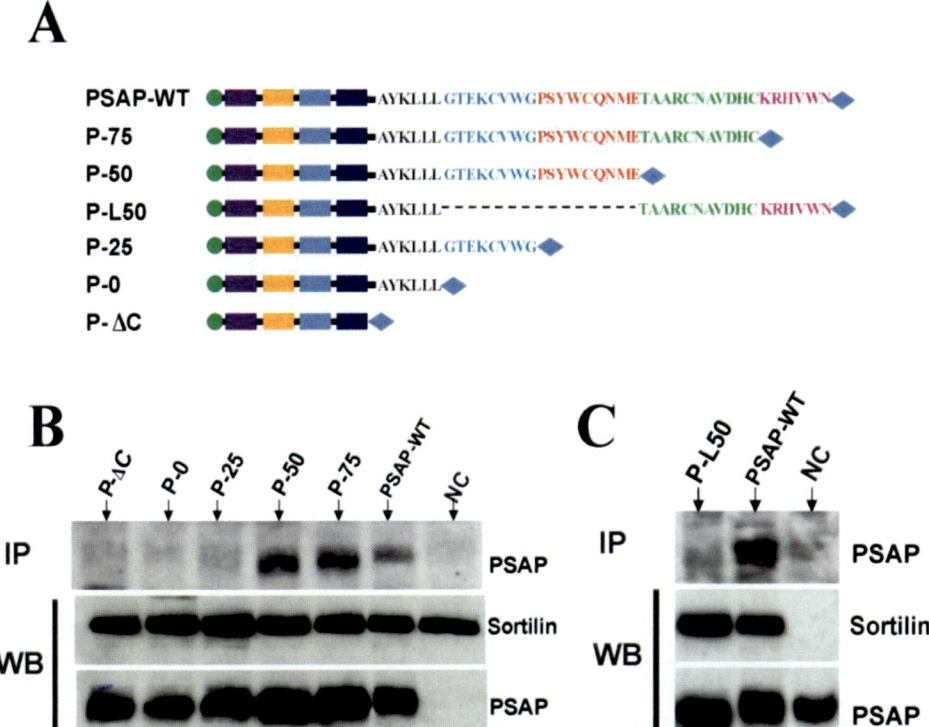

Figure 5. Effects of sequential truncations of the C-terminus of prosaposin. These truncations have been done with the purpose of identifying the putative sortilin binding site in the prosaposin C-terminus, also known as the A-type domain. A) Diagram of truncated prosaposin constructs. PSAP-WT is wild-type prosaposin; P-75 lacks 25% of the A-type domain; P-50, lacks 50% of the A-type domain; P-L50 lacks the first half of the A-type domain, but contains the second half; P-25 lacks 75% of the A-type domain; P-0 contains only linker region; and P-ΔC lacks the whole C-terminus. B) Co-immunoprecipitation with anti-sortilin antibody was conducted to analyze the binding of truncated prosaposin to sortilin. While P-75 and P-50 were precipitated by sortilin, P-0 and P-25 were not. PSAP-WT was used as positive control and P-ΔC as negative control. NC was another negative control, in which the cells were only transfected with the sortilin construct. C) Contrary to PSAP-WT, P-L50 was not pulled-down by sortilin. WB, Western blot as 2% input of sortilin and prosaposin in this experiment. *(Reproduced with permission from Yuan L., Morales C.R. A stretch of 17 Amino Acids in the Prosaposin C-Terminus is Critical for its Binding to Sortilin and Targeting to Lysosomes. J. Histochem. Cytochem. 58: 287-300, 2010)*

gel. Immunoblotting with anti-myc antibody showed that sortilin pulled down the wild type prosaposin, PSAP-WT, and truncated constructs P75 and P50, while it failed to precipitate P-25, P-0 and P-ΔC (Figure 5B). These results demonstrated that the critical domain for the binding of prosaposin to sortilin is located within the first half of the C-terminus of prosaposin (aa524-540). To confirm that the external portion of the C-terminus is not involved in this process we generated a construct termed P-L50, lacking the first half of the A-type domain. In this construct, the external region of the A-type domain was attached to the linker region. As expected, P-L50 was not pulled down by sortilin (Figure 5C) [130]. To verify whether or not the truncations affected the transport of prosaposin to the lysosomes, COS-7 cells were transfected with each prosaposin construct and examined by confocal microscopy after *immunofluorescence labeling*. The wild-type and truncated prosaposin constructs were stained in green with anti-myc antibody. The TGN and lysosomes were stained in red with an anti-Golgin 97 or with an anti-LAMP1 antibody respectively. The nuclei were counterstained blue with Hoechst 33342. The results demonstrated that wild type prosaposin exhibited vesicular and Golgi staining and overlaid with both TGN and lysosomal staining. The distribution of P75 and P50 was similar to that of PSAP-WT (Figure 6A). However, P-ΔC, P0, and P-25 localized only in the perinuclear region and overlaid with anti-Golgin 97, but not with anti-LAMP-1 staining (Figure 6A). Statistical analysis showed that the percentage of overlaid vesicles of P-ΔC, P-0, P-25 and P-L50 significantly decreased as compared to PSAP-WT (P<0.01), while P-50 and P-75 did not (P>0.05) (Figure 6B). In conclusion, the combination of protein truncations has allowed us to identify the putative sortilin binding site in prosaposin. Our findings suggest that this region is located within the first half of the A-type domain on the C-terminus, which is located between aa524 and 540. According to the predicted secondary structure of the C-terminus, this region may contain a β-sheet and several turns, which may be stabilized by proline and tryptophan residues. We have shown that the deletion of this region and the substitution of critical hydrophobic residues abolished the binding of prosaposin to sortilin and the targeting of prosaposin to the lysosomes [130].

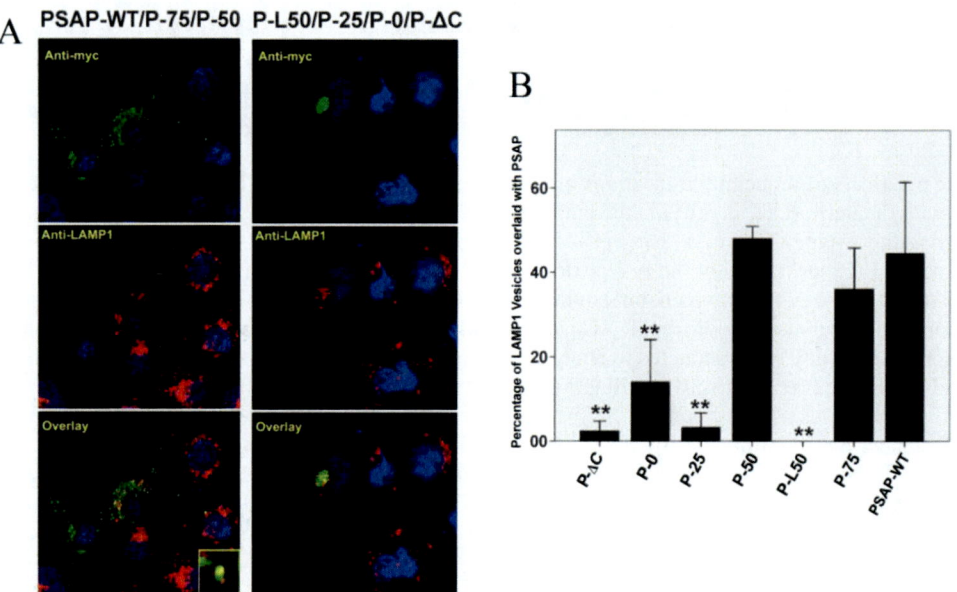

Figure 6. Lysosomal targeting of truncated prosaposin. A) COS-7 cells transfected with truncated prosaposin constructs were examined by confocal microscopy. Prosaposin was stained green with anti-myc antibody, and the TGN or lysosomes stained red with anti-Golgin97 or anti-LAMP-1 antibody. Nuclei appear in blue. Anti-myc staining of cells transfected with PSAP-WT, P-75 and P-50 constructs labeled the perinuclear region and cytoplasmic vesicular structures and overlaid with anti-Golgin 97 and anti-LAMP1 (left panel). P-L50, P-25, P-0 and P-ΔC overlaid only with anti-Golgin staining and labeled the perinuclear region, but not the cytoplasmic vesicular structures (right panel). B) Percentage of LAMP-1 vesicles overlaid with prosaposin truncated constructs (histogram). P-ΔC, P-0, P-25 and P-L50 showed significant decreases in lysosomal transport. P-50 and P-75 revealed no significant difference from PSAP-WT. Error bars indicate ± S.E. and double asterisks represent P<0.01, compared with PSAP-WT. *(Reproduced with permission from Yuan L., Morales C.R. A stretch of 17 Amino Acids in the Prosaposin C-Terminus is Critical for its Binding to Sortilin and Targeting to Lysosomes. J. Histochem. Cytochem. 58: 287-300, 2010)*

V. $G_{M2}AP$ TRAFFICKING

Like prosaposin, the $G_{M2}AP$ interacts with and requires the sortilin sorting receptor to mediate its exit from the TGN and transport to the lysosomal compartment [21]. As noted above, the C-terminal domain of prosaposin, which contains two small α-helices is implicated in the binding of prosaposin to sortilin [15, 16]. Comparison of the C-terminal domain of prosaposin and $G_{M2}AP$ reveals no sequence homology. However, structural analysis demonstrates the presence of a single α-helix in the middle of the $G_{M2}AP$ [17]. In addition, mutations in the α-helix lead to the absence of $G_{M2}AP$ activity in the lysosomes [18]. It was therefore postulated that the α-helix of $G_{M2}AP$ is implicated in its binding to sortilin [17]. To examine this hypothesis, point mutations were introduced to the $G_{M2}AP$ α-helix to change the hydrophobicity, hydrophilicity and charge of selected amino acids. The selection of amino acids to be mutated was based on an analysis of mutations in the α-helix of $G_{M2}AP$ found in patients suffering from the AB variant form of G_{M2} gangliosidosis [18].

The mutants included: M117V, D113A, D113K, D113Y, E123A, E123K, and E123Y (See Table I). The $G_{M2}AP$ mutants were generated in Dr. K. Sandhoff's lab (Kekulé-Institut, Bonn, Germany) and examined by circular dichroism spectroscopy to determine if any conformational changes introduced by the mutations. This structural analysis revealed that certain $G_{M2}AP$ mutants exhibited a normal conformation whereas other presented changes in their tertiary structure. Specifically, the structures of D113A, D113Y, E123A, and E123Y differed from the wild type. It was hypothesized that mutation of amino acids important in the interaction of $G_{M2}AP$ with sortilin, and in particular those causing conformation changes, would result in a reduced ability of $G_{M2}AP$ to bind sortilin.

Table I

Mutants	Mutations and Properties	Comparison with Wild Type Conformation
M117V	Met117 → Val: hydrophobic to hydrophobique	Same
D113A	Asp113 → Ala: hydrophilic acid to hydrophobic	Different
D113K	Asp113 → Lys: hydrophilic acid to hydrophilic basic	Same
D113Y	Asp113 → Tyr: hydrophilic acid to hydrophobic	Different
E123A	Glu123 → Ala: hydrophilic acid to hydrophobic	Different
E123K	Glu123 → Lys: hydrophilic acid to hydrophilic basic	Same
E123Y	Glu123 → Tyr: hydrophilic acid to hydrophobic	Different

The recombinant mutated proteins were translated *in vitro*. Co-IP-assays were performed with these *in vitro* translated proteins and *in vitro* translated sortilin. Using this approach, experiments were performed to examine if the mutant forms of $G_{M2}AP$ could still interact with sortilin. Our *in vitro* Co-IP results showed that E123A and E123Y had a reduced ability to immunoprecipitate sortilin, confirming our predictions based on their structures (Figure 7).

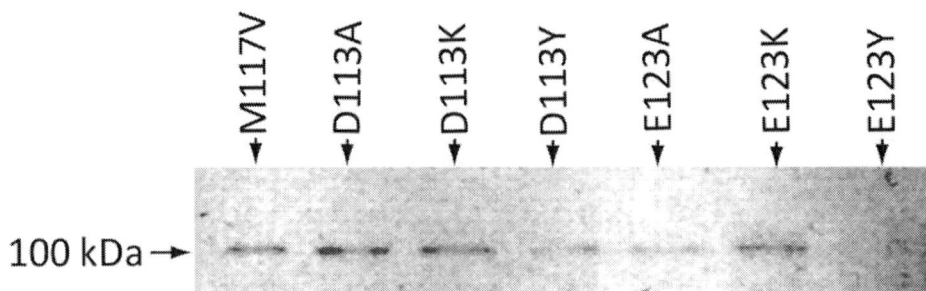

Figure 7. Interaction of $G_{M2}AP$ and sortilin. COS-7 cells were transfected with mutant $G_{M2}AP$ constructs M117V, D113A, D113K, D113Y, E123A, E123K, or E123Y. Co-immunoprecipitation was then performed to test the ability of the mutant constructs to interact with and immunoprecipitate sortilin. All of the constructs tested except E123Y were able to immunoprecipitate sortilin, although to varying degrees.

Both of these mutants represent the replacement of an acidic hydrophilic amino acid residue for a hydrophobic residue. However, when the same amino acid was replaced with a hydrophilic basic one, E123K, the binding to sortilin was unaffected. This suggests that it is the structure of $G_{M2}AP$ and not this specific residue that is important for binding to sortilin. This experiment also demonstrated that when D113 was mutated to D113Y or D113A, there was an altered protein conformation.

Interestingly, while the D113Y mutant inhibited the ability of $G_{M2}AP$ to immunoprecipitate sortilin, the D113A mutant did not. It is possible that the D113Y mutation affected the $G_{M2}AP$ α-helical structure more severely than did the D113A mutation.

VI. THE ROLE OF LIPID RAFTS IN LYSOSOMAL TRAFFICKING

The inhibition of sphingomyelin synthesis with fumonisin B_1 and D609 interferes with the transport of prosaposin to the lysosomes by misrouting this protein to the extracellular space [124].

Given that sphingomyelin and cholesterol are enriched in DRMs, which serve as scaffolds for cellular signaling and trafficking events [133-136], we tested the hypothesis that sortilin and prosaposin reside in DRMs. We also examined whether prosaposin is segregated to DRMs prior to or after its interaction with sortilin. DRMs are defined as membrane microdomains that are insoluble at low temperatures using non-ionic detergents like Triton X-100 [137, 138]. Using sucrose-gradient fractionation, we found that fractions 1 - 8 (10% - 30% sucrose gradient) contained DRMs which floated towards the top of the centrifuge tube. Fractions 9-10 (40% sucrose gradient) possessed soluble proteins and fraction 11 corresponded to the pellet containing insoluble cellular debris [139]. To determine the distribution of sortilin and prosaposin, the resulting fractions were immunoblotted with sortilin and prosaposin antibodies. As a marker for DRMs, the fractions were also blotted with anti-flotillin-1 and 2 antibodies (Figure 8A). Flotillin-1 (47 kDa) was found to be concentrated in fractions 4 and 5, while flotillin-2 was present in DRM fractions 5, 6, 7 as a 42 kDa band. Anti-sortilin (100 kDa) and anti-prosaposin (65 kDa), antibodies yielded bands in DRM fractions 4 through 8 as well as in detergent-soluble fractions 9 and 10 [139].

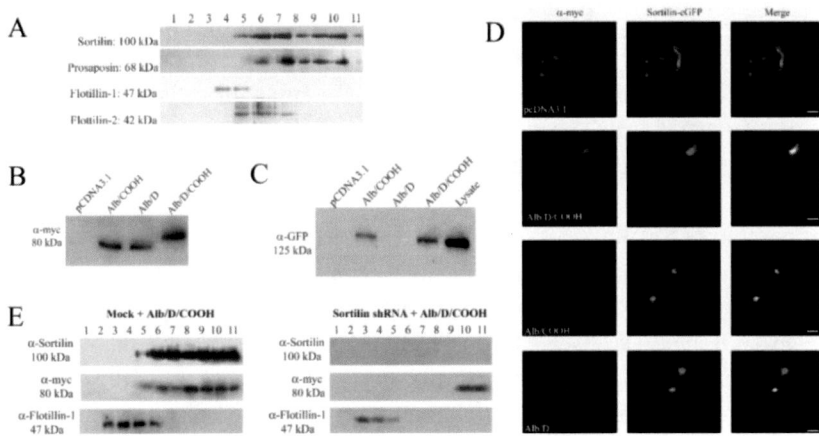

Figure 8. Localization of prosaposin and sortilin to DRMS. A) DRMs were isolated by sucrose gradient fractionation. Fractions 1-8 contain detergent-insoluble membrane fractions, whereas fractions 9 and 10 are the soluble protein fractions. The fractions were immunoblotted with anti-flotillin-1 and 2 antibodies as positive markers of DRMs. Sortilin and its ligand, prosaposin, were present in DRM fractions (5-8) as well as detergent-soluble fractions 9-10. B) COS-7 cells were co-transfected with sortilin-eGFP and the chimeric constructs (Alb/D, Alb/COOH, Alb/D/COOH) or empty pcDNA3.1 vector. Expression of the chimeric constructs was verified by Western blotting. C) The same cells in as in panel B were lysed and immunoprecipitated with an anti-albumin antibody and subsequently immunoblotted with an anti-GFP antibody. Sortilin-eGFP was immunoprecipitated, as a 125 kDa band, by the Alb/D/COOH and Alb/COOH constructs. D) COS-7 cells were co-transfected with similar combinations of constructs as above, fixed, and labeled using anti-myc antibody (red). Cells expressing the Alb/D/COOH construct showed co-localization of the construct with sortilin-eGFP in punctate lysosomal structures, as well as in the perinuclear region (yellow). Alternatively, the Alb/COOH and Alb/D constructs co-localized in yellow with sortilin-eGFP in the perinuclear region, but not in punctate lysosomal structures. E) COS-7 cells that were mock transfected with shRNA or expressing sortilin-specific shRNA were submitted to sucrose-gradient fractionation and the distribution of the Alb/D/COOH construct in the resulting fractions analyzed. In fractions from mock-transfected cells the Alb/D/COOH chimeric construct was present at 80 kDa in both DRM and detergent-soluble fractions. Conversely, in cells expressing sortilin-specific shRNA the chimeric construct was only present in detergent-soluble fractions 10 and 11. *(Reproduced with permission from Canuel M., Bhattacharyya N., Balbis A., Yuan .L, Morales C.R. Sortilin and mannose-6-phosphate receptor sort soluble lysosomal proteins in distinct membrane microdomains. Exp. Cell Res. 315: 240-247, 2009)*

We have reported that the D-domain and the C-terminus of prosaposin, together with sphingomyelin play a crucial role in the transport of prosaposin to the lysosomal compartment [140]. Additionally, the D-domain of prosaposin has been suggested to interact with sphingomyelin [121]. Based on these findings, we postulated that a conditional interaction between the D-domain of prosaposin and sphingomyelin was needed for the eventual binding of the C-terminus of prosaposin to sortilin in DRM platforms in the Golgi apparatus [21, 141]. To test this hypothesis we first examined whether the interaction of prosaposin with sortilin required the D-domain or C-terminus of prosaposin or both [139]. Thus, COS-7 cells were co-transfected with a sortilin construct and one of the following three chimeric constructs: 1) albumin attached to the D-domain and C-terminus of prosaposin (Alb/D/COOH); 2) albumin and the D-domain (Alb/D); 3) albumin and the C-terminus of prosaposin (Alb/COOH). As a negative control, the empty pcDNA3.1 vector was also used. All three of the chimeric constructs produced bands between 70 and 80 kDa when immunoblotted with anti-myc antibody. Lysate from cells transfected with the empty vector did not produce any bands when immunoblotted with anti-myc antibody (Figure 8B).

While sortilin was not immunoprecipitated from the lysate of cells transfected with empty pCDNA3.1 vector or the Alb/D construct, as is seen by the absence of any bands in these lanes, both the Alb/D/COOH and Alb/COOH constructs immunoprecipitated sortilin (Figure 8C). This finding suggests that the C-terminus, but not the D-domain of prosaposin, is required for binding to sortilin.

To investigate the role of the D-domain and C-terminus of prosaposin in the trafficking of prosaposin to the lysosomes, confocal immunomicroscopy was performed. Given the ability of the Alb/D/COOH and Alb/COOH constructs to interact with sortilin, we predicted that these constructs should co-localize with sortilin in lysosomes, whereas the Alb/D construct, which does not interact with sortilin, should be excluded from these structures. As expected, the wild-type Alb/D/COOH was found in the perinuclear Golgi region, as well as in punctuate lysosomal structures (Figure 8D). On the other hand, the Alb/COOH and Alb/D constructs were found to label the perinuclear region exclusively (Figure 8D). Cells transfected with sortilin and an empty pcDNA3.1 vector were examined as a negative control (Figure 8D). These results indicate that while the D-domain of prosaposin is not implicated in the association of prosaposin with sortilin or DRMs, it is required for the exit of prosaposin from the TGN [139]. In addition, our results substantiates our previous observation that the efficient transport of prosaposin to the lysosomes is mediated by both its D-domain and C-terminus [116].

We have also examined whether or not prosaposin is capable of interacting with DRMs without first binding to sortilin or if prosaposin enters the DRM domains only after binding to sortilin [139]. To address these questions, cells stably expressing short hairpin RNA (shRNA) specific to sortilin were transfected with the wild-type Alb/D/COOH construct and the distribution of the proteins in DRMs analyzed. The results of the DRM fractionation were compared to cells expressing Alb/D/COOH, but mock transfected with shRNA.

Examination of the fractions confirmed that in cells expressing sortilin-specific shRNA, there was no sortilin expression, whereas in mock transfected cells sortilin was found in both DRM and detergent-soluble fractions (Figure 8E). Similarly, in mock transfected cells the wild-type Alb/D/COOH construct was found in DRM fractions as well as detergent-soluble fractions (Figure 8E). Conversely, in the absence of sortilin, the wild-type Alb/D/COOH construct was restricted to detergent-soluble fraction 10 (Figure 8E). Under both conditions, flotillin-1 was found in similar DRM fractions, 2-5, (Figure 8E). Thus, the interaction of prosaposin with sortilin is a necessary step for the entry of prosaposin in DRMs and this interaction must occur through the C-terminal domain of prosaposin.

While the full functional significance of DRMs in lysosomal transport remains to be clarified, the high affinity of DRM-localized proteins to lipid rafts suggests that the physical state of the membrane is of functional importance for the proper sorting and/or segregation of molecules [142]. We have provided provide strong evidence that both sortilin and its cargo prosaposin are present in DRMs and that the segregation of prosaposin to DRMs occurs through an interaction of its C-terminus with sortilin.

VII. THE SORTILIN MODEL

We have created a model for the mechanism of sortilin function using prosaposin as guiding protein. We have selected prosaposin because it is the first identified and best studied intracellular ligand of sortilin [21]. The model takes into consideration existing information from our and other laboratories (Figure 9).

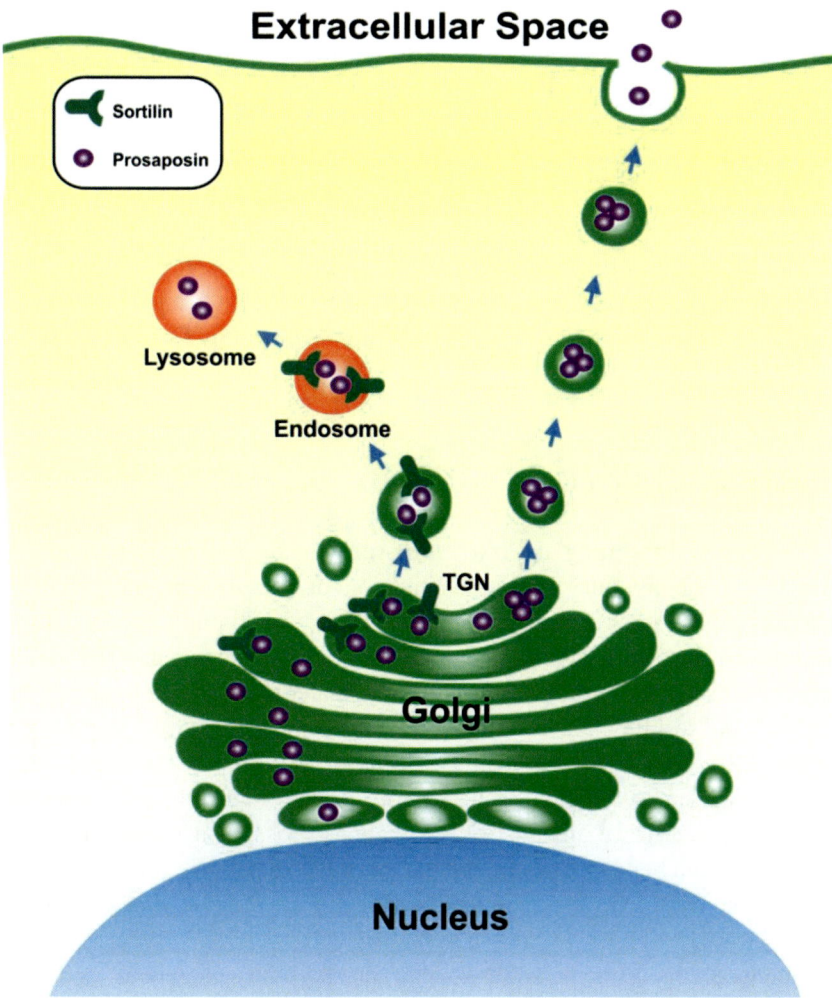

Figure 9. Model showing that the 65 kDa prosaposin must bind sortilin to be transported from TGN to lysosomes. A subset of prosaposin (70 kDa) is fully glycosylated in the TGN and secreted out of the cells.

To be sorted from the Golgi apparatus, sortilin must bind first lysosomal prosaposin (65-68 kDa), which is partially glycosylated. Since a subset of prosaposin is secreted out as a fully glycosylated protein (70 kDa), we postulate that sortilin operates under saturation conditions in the distal mid-saccules or in some distinct, possibly proximal region, of the TGN. In fact treatment with BFA suggests that this is the case, since it leads to the retention of the 65 kDa prosaposin and inhibition of full glycosylation and secretion of the 70 kDa isomer. Next, sortilin must bind one of three ARF1-binding proteins (GGAs 1-3) [21, 82].

Published evidence suggests that the cytoplasmic tail of sortilin also interacts directly with AP-1 and that this protein complex is required for the transport of sortilin and its cargo from the TGN to the endosomal/lysosomal compartments [103]. GGA and AP-1 recruitment to membranes requires the ARF-1 small G protein [143]. ARF-1 is converted from a GDP-bound state to a GTP-bound state by a GEF [144], while GAPs induce hydrolysis of GTP to GDP to deactivate ARF-1 [145].

Recently it has been shown that the GEF, BIG2, may be involved in the recruitment of the multimeric adaptor protein AP-1 to endosomal membrane [146]. However, we have shown that AP-1 is actually recruited to the TGN [103]. Once it has delivered its cargo to the endosomes, sortilin is then recycled back from the endosomes to the TGN via the retromer complex [103].

In spite of recent progress, the mechanism of receptor trafficking is still poorly understood. Investigators working in the field of lysosomal transport suspect that the MPR and sortilin share a "conserved mechanism" that involves sorting and trafficking from the TGN to the endosomal compartment which is not only mediated by GGAs and AP-1, but also by switch proteins.

Inactivation of the Sortilin Gene

Because sortilin is the sorting and trafficking receptor for at least five soluble lysosomal proteins, we tested the hypothesis that inactivation of the sortilin gene would result in lysosomal storage disorders. To our surprise, none of the sortilin nullizygous mice exhibited clinical signs of lysosomal pathologies. Based on this observation we decided to quantify the concentration of prosaposin in the lysosomes of nonciliated epithelial cells lining the efferent ducts of sortilin$^{-/-}$ and sortilin$^{+/+}$ mice.

The rationale for choosing the nonciliated cells of the efferent ducts was based on the following: 1) prosaposin is a major ligand of sortilin [21] and the nonciliated cells of the efferent ducts synthesize large amounts of endogenous prosaposin [115]; 2) the nonciliated cells contain a large number of lysosomes which are well labeled by anti-prosaposin allowing its easy quantification [115]; 3) the nonciliated cells are known to endocytose luminal prosaposin [114, 147].

In consequence, the nonciliated cells can target both exogenous and endogenous prosaposin to the lysosomes, and therefore, these cells constituted an excellent *in vivo* system to analyze the effect of sortilin gene inactivation in cells deprived or not of luminal prosaposin.

First we examined the effect of sortilin inactivation on the level of prosaposin in endosomes and lysosomes in nonciliated cells of mouse efferent ducts using electron microscopy and immunogold labeling with an anti-prosaposin antibody. Our results demonstrated a 40% reduction in the level of prosaposin in endosomes of nonciliated cells of sortilin$^{-/-}$ mice as compared to WT mice (Table 2). Similarly, we observed a 15% reduction in gold labeling of prosaposin in the lysosomes of sortilin$^{-/-}$ mice (Table II). The differences between sortilin$^{-/-}$ and sortilin$^{+/+}$ mice were statistically significant (p≤0.05).

Prosaposin in the luminal fluid of the efferent ducts originates from testicular Sertoli cells and is endocytosed by the epithelial cells of the efferent ducts by an unknown mechanism that may or may not involve sortilin. Consequently, ligation of the efferent ducts of sortilin$^{+/+}$ and sortilin$^{-/-}$ constituted an exciting experiment to determine the luminal contribution of prosaposin to the lysosomes. In addition, this experiment allowed us to compare the effect of ligation on the labeling of endosomes and lysosomes in sortilin$^{+/+}$ and sortilin$^{-/-}$ mice.

Our results demonstrated a 90% reduction of prosaposin labeling in the endosomes of sortilin$^{+/+}$ mice after ligation (Table II) and a 76% reduction of prosaposin labeling in the endosomes of prosaposin$^{-/-}$ mice (Table II). We observed a 25% reduction in prosaposin labeling in the lysosomes of WT mice after ligation and a 22% reduction in prosaposin labeling in the lysosomes of prosaposin$^{-/-}$ mice after ligation (Table II).

The differences between ligated and non-ligated mice were statistically significant ($p \leq 0.05$). Prosaposin labeling of endosomes of non-ligated sortilin+/+ and sortilin-/- cells revealed no significant changes (Table II).

Table 2. Immunogold labeling of lysosomes and endosomes

Compartments	WNL	WL	KONL	KOL
Lysosomes	86±5	92±3	329±2	259±2
Endosomes	1±1	8±3	37±1	9±3

WNL, wild type non ligated; WL, wild type ligated; KONL, KO non-ligated; KOL, KO ligated. The values are expressed as number of gold particles per 10^6 pixels ± SEM.

Thus, our results demonstrated that the inactivation produced a partial but significant reduction in the level of prosaposin in endosomes and lysosomes. Since the nonciliated cells are known to endocytose luminal prosaposin produced by the testis [114, 147], we decided to exclude this protein from the lumen of the efferent ducts by surgical ligation. The ligation of the efferent ducts resulted in a significant reduction of prosaposin labeling in endosomes and lysosomes of both WT and nullizygous mice. This observation was not surprising since prosaposin has been previously shown to be internalized and delivered to the lysosomal compartment via the LRP receptor [148].

In conclusion, our results validate the role assigned to sortilin as a sorting receptor and suggest the existence of an alternative pathway possibly mediated by another member of the VPS10 family, such as SorLA, SorCS1, SorCS2 and/or SorCS3. Some members of the VPS10 receptor family, however, may have intracellular domains containing imperfect sequences essential for lysosomal trafficking [149]. A compelling experiment would be the generation of double or triple knockouts for sortilin, SorLA and/or LRP.

VIII. SIGNIFICANCE OF THIS STUDY

Over the past ten years we have disclosed that the prosaposin C-terminus contains the binding site to sortilin which is located in the first half of this region (aa 524-540) [116, 130]. Using an in-silico based approach we have built a molecular 3D structural model of the binding site within the prosaposin C-terminus. This technique allowed us to propose a molecular model that explains the formation of the prosaposin-sortilin complex. In the near future we also hope to identify the important amino acid residues in the sortilin receptor engaged in the recognition and binding of prosaposin. Eventually, the role of these residues will be verified by site-directed mutagenesis. The molecular structure of the prosaposin-sortilin complex will be also built in conjunction with Cryo-EM or X-ray crystallography.

Our laboratory also demonstrated that the 65 kDa prosaposin is partially glycosylated and transported to the lysosomes while the fully glycosylated 70 kDa prosaposin is secreted to the extracellular space [116, 118, 148]. The elucidation of the mechanisms that lead to different glycosylation forms and diverse trafficking are compelling. We have proposed a sortilin saturation model to explain why partially glycosylated prosaposin (65 kDa) escapes the Golgi via the sortilin pathway (Figure 9). According this model, sortilin binds only partially glycosylated prosaposin under saturation conditions before it reaches the TGN. On the other hand, the remaining unbound molecules of prosaposin flow to the TGN where they are fully glycosylated to a 70 kDa form. The 70 kDa prosaposin possibly aggregates in the TGN and is secreted via a regulated mechanism. The model also proposes that prosaposin bound to sortilin escapes full glycosylation because the receptor hinders the activator protein from the glycosylation enzymes. This explanation is supported by two reports on the composition of the oligosaccharide side chains on the four different saposin molecules, which represent the glycosylation status of the 65 kDa precursor. Indeed, saposin A, B and C have been found to contain mainly complex oligosaccharide chains while saposin D contains only oligomannose-type oligosaccharides [150-154]. Since the saposin D domain is topologically closest to sortilin in the prosaposin-sortilin complex, as compared to the saposin A, B, and C domains, this receptor may be responsible for hampering the glycosylation of the saposin D domain.

We are also interested in determining whether the 70 kDa prosaposin is endocytosed on the cell surface and transported to the lysosomes to yield mature saposins, in a similar fashion to the classical processing of prosaposin delivered from the Golgi apparatus [147, 155]. Preliminary evidence from two laboratories suggests that prosaposin may be internalized to the lysosomes via the LRPs 1 and 2 [148, 156]. In addition to LRPs, the MPR may also play a minor role in the endocytosis of prosaposin [148].

The disclosure of this novel mechanism of sorting and trafficking of lysosomal proteins independent of the MPR pathway, and the identification of the prosaposin endosomal/lysosomal targeting motif may have great significance in the treatment of lysosomal storage disorders. We have demonstrated that albumin linked to the D-domain and C-terminus of prosaposin can be effectively directed to the lysosomes via the sortilin receptor. Thus, this sequence may also be used to target hydrolases and biologically active proteins to the lysosomes to restore defective proteins and/or to bypass the canonical pathway.

Furthermore, a number of life threatening pathogens such as the hepatitis C virus [157], HIV [158], *Trypanosome cruzi* [158], *Mycobacterium tuberculosis* [159, 160], *Mycobacterium leprae* [160, 161] for which there are no effective treatments, enter the cell via endocytosis and through the endosomal/lysosomal compartment. The engineering of biologically active proteins against these pathogens can be effectively targeted to the endosomal/lysosomal compartment by linking them to prosaposin targeting sequence.

REFERENCES

[1] Brocks, J.J., et al., Archean molecular fossils and the early rise of eukaryotes. *Science*, 1999. 285(5430): p. 1033-6.

[2] Griffiths, G. and K. Simons, The trans Golgi network: sorting at the exit site of the Golgi complex. *Science*, 1986. 234(4775): p. 438-43.

[3] Tekirian, T.L., The central role of the trans-Golgi network as a gateway of the early secretory pathway: physiologic vs nonphysiologic protein transit. *Exp. Cell Res*, 2002. 281(1): p. 9-18.

[4] Lobel, P., et al., Mutations in the cytoplasmic domain of the 275 kd mannose 6-phosphate receptor differentially alter lysosomal enzyme sorting and endocytosis. *Cell*, 1989. 57(5): p. 787-96.

[5] Stein, M., et al., 46-kDa mannose 6-phosphate-specific receptor: purification, subunit composition, chemical modification. *Biol. Chem. Hoppe. Seyler*, 1987. 368(8): p. 927-36.

[6] Lobel, P., et al., Cloning of the bovine 215-kDa cation-independent mannose 6-phosphate receptor. *Proc. Natl. Acad. Sci. U S A*, 1987. 84(8): p. 2233-7.

[7] Dahms, N.M., et al., 46 kd mannose 6-phosphate receptor: cloning, expression, and homology to the 215 kd mannose 6-phosphate receptor. *Cell*, 1987. 50(2): p. 181-92.

[8] Hancock, M.K., et al., Identification of residues essential for carbohydrate recognition by the insulin-like growth factor II/mannose 6-phosphate receptor. *J. Biol. Chem*, 2002. 277(13): p. 11255-64.

[9] Kornfeld, S., Structure and function of the mannose 6-phosphate/insulinlike growth factor II receptors. *Annu. Rev. Biochem*, 1992. 61: p. 307-30.

[10] Nadimpalli, S.K. and K. von Figura, Identification of the putative mannose 6-phosphate receptor (MPR 46) protein in the invertebrate mollusc. *Biosci. Rep*, 2002. 22(5-6): p. 513-21.

[11] Nadimpalli, S.K., et al., Mannose 6-phosphate receptors (MPR 300 and MPR 46) from a teleostean fish (trout). *Comp. Biochem. Physiol. B. Biochem. Mol. Biol*, 1999. 123(3): p. 261-5.

[12] Matzner, U., et al., Expression of mannose 6-phosphate receptors in chicken. *Dev. Dyn*, 1996. 207(1): p. 11-24.

[13] Nolan, C.M., et al., Mannose 6-phosphate receptors in an ancient vertebrate, zebrafish. *Dev. Genes Evol*, 2006. 216(3): p. 144-51.

[14] Palmer, D., Prehistoric Past Revealed: The Four Billion Year History of Life on Earth. 2003: University of California Press 176.

[15] Marcusson, E.G., et al., The sorting receptor for yeast vacuolar carboxypeptidase Y is encoded by the VPS10 gene. *Cell*, 1994. 77(4): p. 579-86.

[16] Kirsch, T., et al., Purification and initial characterization of a potential plant vacuolar targeting receptor. *Proc. Natl. Acad. Sci. U S A*, 1994. 91(8): p. 3403-7.

[17] Westergaard, U.B., et al., Functional organization of the sortilin Vps10p domain. *J. Biol. Chem*, 2004. 279(48): p. 50221-9.

[18] Hermey, G., et al., SorCS1, a member of the novel sorting receptor family, is localized in somata and dendrites of neurons throughout the murine brain. *Neurosci. Lett*, 2001. 313(1-2): p. 83-7.

[19] Jacobsen, L., et al., Activation and functional characterization of the mosaic receptor SorLA/LR11. *J. Biol. Chem*, 2001. 276(25): p. 22788-96.

[20] Hermans-Borgmeyer, I., et al., Expression of the 100-kDa neurotensin receptor sortilin during mouse embryonal development. *Brain. Res. Mol. Brain Res*, 1999. 65(2): p. 216-9.

[21] Lefrancois, S., et al., The lysosomal trafficking of sphingolipid activator proteins (SAPs) is mediated by sortilin. *Embo J*, 2003. 22(24): p. 6430-7.

[22] Mazella, J., et al., The 100-kDa neurotensin receptor is gp95/sortilin, a non-G-protein-coupled receptor. *J. Biol. Chem*, 1998. 273(41): p. 26273-6.

[23] Nielsen, M.S., et al., Sortilin/neurotensin receptor-3 binds and mediates degradation of lipoprotein lipase. *J. Biol. Chem*, 1999. 274(13): p. 8832-6.

[24] Petersen, C.M., et al., Molecular identification of a novel candidate sorting receptor purified from human brain by receptor-associated protein affinity chromatography. *J. Biol. Chem*, 1997. 272(6): p. 3599-605.

[25] Hampe, W., et al., The genes for the human VPS10 domain-containing receptors are large and contain many small exons. *Hum. Genet*, 2001. 108(6): p. 529-36.

[26] Zeng, J., A.J. Hassan, and C.R. Morales, Study of the mouse sortilin gene: Effects of its transient silencing by RNA interference in TM4 Sertoli cells. *Mol. Reprod. Dev*, 2004. 68(4): p. 469-75.

[27] Steinman, R.M., et al., Endocytosis and the recycling of plasma membrane. *J. Cell Biol*, 1983. 96(1): p. 1-27.

[28] Sandhoff, K. and T. Kolter, Biosynthesis and degradation of mammalian glycosphingolipids. *Philos. Trans. R. Soc. Lond. B. Biol. Sci*, 2003. 358(1433): p. 847-61.

[29] Luzio, J.P., P.R. Pryor, and N.A. Bright, Lysosomes: fusion and function. *Nat. Rev. Mol. Cell Biol*, 2007. 8(8): p. 622-32.

[30] Peters, C. and K. von Figura, Biogenesis of lysosomal membranes. *FEBS Lett*, 1994. 346(1): p. 108-14.

[31] Carlsson, S.R., et al., Isolation and characterization of human lysosomal membrane glycoproteins, h-lamp-1 and h-lamp-2. Major sialoglycoproteins carrying polylactosaminoglycan. *J. Biol. Chem*, 1988. 263(35): p. 18911-9.

[32] Schulze, H., T. Kolter, and K. Sandhoff, Principles of lysosomal membrane degradation Cellular topology and biochemistry of lysosomal lipid degradation. *Biochim. Biophys. Acta*, 2008.

[33] Wilkening, G., T. Linke, and K. Sandhoff, Lysosomal degradation on vesicular membrane surfaces. Enhanced glucosylceramide degradation by lysosomal anionic lipids and activators. *J. Biol. Chem*, 1998. 273(46): p. 30271-8.

[34] Umebayashi, K., The roles of ubiquitin and lipids in protein sorting along the endocytic pathway. *Cell Struct. Funct*, 2003. 28(5): p. 443-53.

[35] Frederick, T.E., et al., Bis(monoacylglycero)phosphate forms stable small lamellar vesicle structures: insights into vesicular body formation in endosomes. *Biophys. J*, 2009. 96(5): p. 1847-55.

[36] Kolter, T. and K. Sandhoff, Principles of lysosomal membrane digestion: stimulation of sphingolipid degradation by sphingolipid activator proteins and anionic lysosomal lipids. *Annu. Rev. Cell Dev. Biol*, 2005. 21: p. 81-103.

[37] Meier, E.M., et al., The human GM2 activator protein. A substrate specific cofactor of beta-hexosaminidase A. *J. Biol. Chem*, 1991. 266(3): p. 1879-87.

[38] Wright, C.S., S.C. Li, and F. Rastinejad, Crystal structure of human GM2-activator protein with a novel beta-cup topology. *J. Mol. Biol*, 2000. 304(3): p. 411-22.

[39] Furst, W. and K. Sandhoff, Activator proteins and topology of lysosomal sphingolipid catabolism. *Biochim. Biophys. Acta*, 1992. 1126(1): p. 1-16.

[40] Hechtman, P. and D. LeBlanc, Purification and properties of the hexosaminidase A-activating protein from human liver. *Biochem. J*, 1977. 167(3): p. 693-701.

[41] Morimoto, S., et al., Interaction of saposins, acidic lipids, and glucosylceramidase. *J. Biol. Chem*, 1990. 265(4): p. 1933-7.

[42] Sandhoff, K., The GM2-gangliosidoses and the elucidation of the beta-hexosaminidase system. *Adv. Genet*, 2001. 44: p. 67-91.

[43] Burg, J., et al., Mapping of the gene coding for the human GM2 activator protein to chromosome 5. *Ann. Hum. Genet*, 1985. 49(Pt 1): p. 41-5.

[44] Conzelmann, E. and K. Sandhoff, Purification and characterization of an activator protein for the degradation of glycolipids GM2 and GA2 by hexosaminidase A. *Hoppe. Seylers. Z. Physiol. Chem*, 1979. 360(12): p. 1837-49.

[45] Hama, Y., Y.T. Li, and S.C. Li, Interaction of GM2 activator protein with glycosphingolipids. *J. Biol. Chem*, 1997. 272(5): p. 2828-33.

[46] Schuette, C.G., et al., Sphingolipid activator proteins: proteins with complex functions in lipid degradation and skin biogenesis. *Glycobiology*, 2001. 11(6): p. 81R-90R.

[47] Ahn, V.E., et al., Crystal structure of saposin B reveals a dimeric shell for lipid binding. *Proc. Natl. Acad. Sci. U S A*, 2003. 100(1): p. 38-43.

[48] Kishimoto, Y., M. Hiraiwa, and J.S. O'Brien, Saposins: structure, function, distribution, and molecular genetics. *J. Lipid. Res*, 1992. 33(9): p. 1255-67.

[49] Zhao, Q., et al., Mouse testicular sulfated glycoprotein-1: sequence analysis of the common backbone structure of prosaposins. *J. Androl*, 1998. 19(2): p. 165-74.

[50] Harzer, K., M. Hiraiwa, and B.C. Paton, Saposins (sap) A and C activate the degradation of galactosylsphingosine. *FEBS Lett*, 2001. 508(1): p. 107-10.

[51] Ho, M.W. and J.S. O'Brien, Gaucher's disease: deficiency of 'acid' -glucosidase and reconstitution of enzyme activity in vitro. *Proc. Natl. Acad. Sci. U S A*, 1971. 68(11): p. 2810-3.

[52] Klein, A., et al., Sphingolipid activator protein D (sap-D) stimulates the lysosomal degradation of ceramide in vivo. *Biochem. Biophys. Res. Commun*, 1994. 200(3): p. 1440-8.

[53] Kretz, K.A., et al., Characterization of a mutation in a family with saposin B deficiency: a glycosylation site defect. *Proc. Natl. Acad. Sci. U S A*, 1990. 87(7): p. 2541-4.

[54] Matsuda, J., et al., Mutation in saposin D domain of sphingolipid activator protein gene causes urinary system defects and cerebellar Purkinje cell degeneration with accumulation of hydroxy fatty acid-containing ceramide in mouse. *Hum. Mol. Genet*, 2004. 13(21): p. 2709-23.

[55] Spiegel, R., et al., A mutation in the saposin A coding region of the prosaposin gene in an infant presenting as Krabbe disease: first report of saposin A deficiency in humans. *Mol. Genet. Metab*, 2005. 84(2): p. 160-6.

[56] Tylki-Szymanska, A., et al., Non-neuronopathic Gaucher disease due to saposin C deficiency. *Clin. Genet*, 2007. 72(6): p. 538-42.

[57] Conzelmann, E. and K. Sandhoff, AB variant of infantile GM2 gangliosidosis: deficiency of a factor necessary for stimulation of hexosaminidase A-catalyzed degradation of ganglioside GM2 and glycolipid GA2. *Proc. Natl. Acad. Sci. U S A*, 1978. 75(8): p. 3979-83.

[58] Griffiths, G.M. and S. Isaaz, Granzymes A and B are targeted to the lytic granules of lymphocytes by the mannose-6-phosphate receptor. *J. Cell Biol*, 1993. 120(4): p. 885-96.

[59] Nguyen, G., Renin/prorenin receptors. *Kidney Int*, 2006. 69(9): p. 1503-6.

[60] MacDonald, R.G., et al., A single receptor binds both insulin-like growth factor II and mannose-6-phosphate. *Science*, 1988. 239(4844): p. 1134-7.

[61] Lee, S.J. and D. Nathans, Proliferin secreted by cultured cells binds to mannose 6-phosphate receptors. *J. Biol. Chem*, 1988. 263(7): p. 3521-7.

[62] Kang, J.X., Y. Li, and A. Leaf, Mannose-6-phosphate/insulin-like growth factor-II receptor is a receptor for retinoic acid. *Proc. Natl. Acad. Sci. U S A*, 1997. 94(25): p. 13671-6.

[63] Nykjaer, A., et al., Mannose 6-phosphate/insulin-like growth factor-II receptor targets the urokinase receptor to lysosomes via a novel binding interaction. *J. Cell Biol*, 1998. 141(3): p. 815-28.

[64] Oesterreicher, S., et al., Interaction of insulin-like growth factor II (IGF-II) with multiple plasma proteins: high affinity binding of plasminogen to IGF-II and IGF-binding protein-3. *J. Biol. Chem*, 2005. 280(11): p. 9994-10000.

[65] Tong, P.Y. and S. Kornfeld, Ligand interactions of the cation-dependent mannose 6-phosphate receptor. Comparison with the cation-independent mannose 6-phosphate receptor. *J. Biol. Chem*, 1989. 264(14): p. 7970-5.

[66] Ludwig, T., et al., Differential sorting of lysosomal enzymes in mannose 6-phosphate receptor-deficient fibroblasts. *Embo J*, 1994. 13(15): p. 3430-7.

[67] Reitman, M.L., A. Varki, and S. Kornfeld, Fibroblasts from patients with I-cell disease and pseudo-Hurler polydystrophy are deficient in uridine 5'-diphosphate-N-acetylglucosamine: glycoprotein N-acetylglu cosaminylphosphotransferase activity. *J. Clin. Invest*, 1981. 67(5): p. 1574-9.

[68] Varki, A.P., M.L. Reitman, and S. Kornfeld, Identification of a variant of mucolipidosis III (pseudo-Hurler polydystrophy): a catalytically active N-acetylglucosaminylphosphotransferase that fails to phosphorylate lysosomal enzymes. *Proc. Natl. Acad. Sci. U S A*, 1981. 78(12): p. 7773-7.

[69] Waheed, A., et al., Deficiency of UDP-N-acetylglucosamine:lysosomal enzyme N-acetylglucosamine-1-phosphotransferase in organs of I-cell patients. *Biochem. Biophys. Res. Commun*, 1982. 105(3): p. 1052-8.

[70] Hasilik, A., A. Waheed, and K. von Figura, Enzymatic phosphorylation of lysosomal enzymes in the presence of UDP-N-acetylglucosamine. Absence of the activity in I-cell fibroblasts. *Biochem. Biophys. Res. Commun*, 1981. 98(3): p. 761-7.

[71] Kornfeld, S., Trafficking of lysosomal enzymes in normal and disease states. *J. Clin. Invest*, 1986. 77(1): p. 1-6.

[72] Varki, A. and S. Kornfeld, Identification of a rat liver alpha-N-acetylglucosaminyl phosphodiesterase capable of removing "blocking" alpha-N-acetylglucosamine residues from phosphorylated high mannose oligosaccharides of lysosomal enzymes. *J. Biol. Chem*, 1980. 255(18): p. 8398-401.

[73] von Figura, K. and A. Hasilik, Lysosomal enzymes and their receptors. *Annu. Rev. Biochem*, 1986. 55: p. 167-93.

[74] Rijnboutt, S., et al., Mannose 6-phosphate-independent membrane association of cathepsin D, glucocerebrosidase, and sphingolipid-activating protein in HepG2 cells. *J. Biol. Chem*, 1991. 266(8): p. 4862-8.

[75] Tanaka, Y., R. Tanaka, and M. Himeno, Lysosomal cysteine protease, cathepsin H, is targeted to lysosomes by the mannose 6-phosphate-independent system in rat hepatocytes. *Biol. Pharm. Bull*, 2000. 23(7): p. 805-9.

[76] Gottesman, M.M., Transformation-dependent secretion of a low molecular weight protein by murine fibroblasts. *Proc. Natl. Acad. Sci. U S A*, 1978. 75(6): p. 2767-71.

[77] Johnson, L.M., V.A. Bankaitis, and S.D. Emr, Distinct sequence determinants direct intracellular sorting and modification of a yeast vacuolar protease. *Cell*, 1987. 48(5): p. 875-85.

[78] Valls, L.A., et al., Protein sorting in yeast: the localization determinant of yeast vacuolar carboxypeptidase Y resides in the propeptide. *Cell*, 1987. 48(5): p. 887-97.

[79] Claveau, D. and D. Riendeau, Mutations of the C-terminal end of cathepsin K affect proenzyme secretion and intracellular maturation. Biochem. *Biophys. Res. Commun*, 2001. 281(2): p. 551-7.

[80] Canuel, M., et al., Sortilin mediates the lysosomal targeting of cathepsins D and H. *Biochem. Biophys. Res. Commun*, 2008. 373(2): p. 292-7.

[81] Ni, X. and C.R. Morales, The lysosomal trafficking of acid sphingomyelinase is mediated by sortilin and mannose 6-phosphate receptor. *Traffic*, 2006. 7(7): p. 889-902.

[82] Nielsen, M.S., et al., The sortilin cytoplasmic tail conveys Golgi-endosome transport and binds the VHS domain of the GGA2 sorting protein. *Embo J*, 2001. 20(9): p. 2180-90.

[83] Takatsu, H., et al., Golgi-localizing, gamma-adaptin ear homology domain, ADP-ribosylation factor-binding (GGA) proteins interact with acidic dileucine sequences within the cytoplasmic domains of sorting receptors through their Vps27p/Hrs/STAM (VHS) domains. *J. Biol. Chem*, 2001. 276(30): p. 28541-5.

[84] Puertollano, R., et al., Sorting of mannose 6-phosphate receptors mediated by the GGAs. *Science*, 2001. 292(5522): p. 1712-6.

[85] Traub, L.M., Common principles in clathrin-mediated sorting at the Golgi and the plasma membrane. *Biochim. Biophys. Acta*, 2005. 1744(3): p. 415-37.

[86] Hirst, J., M.R. Lindsay, and M.S. Robinson, GGAs: roles of the different domains and comparison with AP-1 and clathrin. *Mol. Biol. Cell*, 2001. 12(11): p. 3573-88.

[87] Dell'Angelica, E.C., et al., GGAs: a family of ADP ribosylation factor-binding proteins related to adaptors and associated with the Golgi complex. *J. Cell Biol*, 2000. 149(1): p. 81-94.

[88] Robinson, M.S., Adaptable adaptors for coated vesicles. *Trends Cell Biol*, 2004. 14(4): p. 167-74.

[89] Hirst, J., et al., A family of proteins with gamma-adaptin and VHS domains that facilitate trafficking between the trans-Golgi network and the vacuole/lysosome. *J. Cell Biol*, 2000. 149(1): p. 67-80.

[90] Boman, A.L., et al., A family of ADP-ribosylation factor effectors that can alter membrane transport through the trans-Golgi. *Mol. Biol. Cell*, 2000. 11(4): p. 1241-55.

[91] Ghosh, P. and S. Kornfeld, The GGA proteins: key players in protein sorting at the trans-Golgi network. *Eur. J. Cell Biol*, 2004. 83(6): p. 257-62.

[92] Misra, S., B.M. Beach, and J.H. Hurley, Structure of the VHS domain of human Tom1 (target of myb 1): insights into interactions with proteins and membranes. *Biochemistry*, 2000. 39(37): p. 11282-90.

[93] Collins, B.M., P.J. Watson, and D.J. Owen, The structure of the GGA1-GAT domain reveals the molecular basis for ARF binding and membrane association of GGAs. *Dev. Cell*, 2003. 4(3): p. 321-32.

[94] Bonifacino, J.S., The GGA proteins: adaptors on the move. *Nat. Rev. Mol. Cell Biol*, 2004. 5(1): p. 23-32.

[95] Lefrancois, S. and P.J. McCormick, The Arf GEF GBF1 is required for GGA recruitment to Golgi membranes. *Traffic*, 2007. 8(10): p. 1440-51.

[96] Manolea, F., et al., Distinct functions for Arf guanine nucleotide exchange factors at the Golgi complex: GBF1 and BIGs are required for assembly and maintenance of the Golgi stack and trans-Golgi network, respectively. *Mol. Biol. Cell*, 2008. 19(2): p. 523-35.

[97] Zhao, X., et al., GBF1, a cis-Golgi and VTCs-localized ARF-GEF, is implicated in ER-to-Golgi protein traffic. *J. Cell Sci*, 2006. 119(Pt 18): p. 3743-53.

[98] Zhao, X., T.K. Lasell, and P. Melancon, Localization of large ADP-ribosylation factor-guanine nucleotide exchange factors to different Golgi compartments: evidence for distinct functions in protein traffic. *Mol. Biol. Cell*, 2002. 13(1): p. 119-33.

[99] Robinson, M.S. and J.S. Bonifacino, Adaptor-related proteins. *Curr. Opin. Cell Biol*, 2001. 13(4): p. 444-53.

[100] Meyer, C., et al., mu1A-adaptin-deficient mice: lethality, loss of AP-1 binding and rerouting of mannose 6-phosphate receptors. *Embo J*, 2000. 19(10): p. 2193-203.

[101] Ohno, H., et al., Interaction of tyrosine-based sorting signals with clathrin-associated proteins. *Science*, 1995. 269(5232): p. 1872-5.

[102] Owen, D.J. and P.R. Evans, A structural explanation for the recognition of tyrosine-based endocytotic signals. *Science*, 1998. 282(5392): p. 1327-32.

[103] Canuel, M., et al., AP-1 and retromer play opposite roles in the trafficking of sortilin between the Golgi apparatus and the lysosomes. *Biochem. Biophys. Res. Commun*, 2008. 366(3): p. 724-30.

[104] Doray, B., et al., Interaction of the cation-dependent mannose 6-phosphate receptor with GGA proteins. *J. Biol. Chem*, 2002. 277(21): p. 18477-82.

[105] Meyer, C., et al., Mu 1A deficiency induces a profound increase in MPR300/IGF-II receptor internalization rate. *J. Cell Sci*, 2001. 114(Pt 24): p. 4469-76.

[106] Seaman, M.N., J.M. McCaffery, and S.D. Emr, A membrane coat complex essential for endosome-to-Golgi retrograde transport in yeast. *J. Cell Biol*, 1998. 142(3): p. 665-81.

[107] Seaman, M.N., Identification of a novel conserved sorting motif required for retromer-mediated endosome-to-TGN retrieval. *J. Cell Sci*, 2007. 120(Pt 14): p. 2378-89.

[108] Arighi, C.N., et al., Role of the mammalian retromer in sorting of the cation-independent mannose 6-phosphate receptor. *J. Cell Biol*, 2004. 165(1): p. 123-33.

[109] Haft, C.R., et al., Human orthologs of yeast vacuolar protein sorting proteins Vps26, 29, and 35: assembly into multimeric complexes. *Mol. Biol. Cell*, 2000. 11(12): p. 4105-16.

[110] Collins, B.M., et al., Vps29 has a phosphoesterase fold that acts as a protein interaction scaffold for retromer assembly. *Nat. Struct. Mol. Biol*, 2005. 12(7): p. 594-602.

[111] Damen, E., et al., The human Vps29 retromer component is a metallo-phosphoesterase for a cation-independent mannose 6-phosphate receptor substrate peptide. *Biochem. J*, 2006. 398(3): p. 399-409.

[112] Saint-Pol, A., et al., Clathrin adaptor epsinR is required for retrograde sorting on early endosomal membranes. *Dev. Cell*, 2004. 6(4): p. 525-38.

[113] Scott, G.K., et al., A PACS-1, GGA3 and CK2 complex regulates CI-MPR trafficking. *Embo J*, 2006. 25(19): p. 4423-35.

[114] Igdoura, S.A. and C.R. Morales, Role of sulfated glycoprotein-1 (SGP-1) in the disposal of residual bodies by Sertoli cells of the rat. *Mol. Reprod. Dev*, 1995. 40(1): p. 91-102.

[115] Rosenthal, A.L., et al., Hormonal regulation of sulfated glycoprotein-1 synthesis by nonciliated cells of the efferent ducts of adult rats. *Mol. Reprod. Dev,* 1995. 40(1): p. 69-83.

[116] Zhao, Q. and C.R. Morales, Identification of a novel sequence involved in lysosomal sorting of the sphingolipid activator protein prosaposin. *J. Biol. Chem*, 2000. 275(32): p. 24829-39.

[117] O'Brien, J.S., et al., Identification of the neurotrophic factor sequence of prosaposin. *Faseb. J*, 1995. 9(8): p. 681-5.

[118] Igdoura, S.A., A. Rasky, and C.R. Morales, Trafficking of sulfated glycoprotein-1 (prosaposin) to lysosomes or to the extracellular space in rat Sertoli cells. *Cell Tissue Res*, 1996. 283(3): p. 385-94.

[119] Chanat, E. and W.B. Huttner, Milieu-induced, selective aggregation of regulated secretory proteins in the trans-Golgi network. *J. Cell Biol,* 1991. 115(6): p. 1505-19.

[120] Morimoto, S., et al., Saposin A: second cerebrosidase activator protein. *Proc. Natl. Acad. Sci. U S A*, 1989. 86(9): p. 3389-93.

[121] Morimoto, S., et al., Saposin D: a sphingomyelinase activator. *Biochem. Biophys. Res. Commun*, 1988. 156(1): p. 403-10.

[122] O'Brien, J.S. and Y. Kishimoto, Saposin proteins: structure, function, and role in human lysosomal storage disorders. *Faseb. J*, 1991. 5(3): p. 301-8.

[123] Vielhaber, G., R. Hurwitz, and K. Sandhoff, Biosynthesis, processing, and targeting of sphingolipid activator protein (SAP)precursor in cultured human fibroblasts. Mannose 6-phosphate receptor-independent endocytosis of SAP precursor. *J. Biol. Chem*, 1996. 271(50): p. 32438-46.

[124] Lefrancois, S., et al., Role of sphingolipids in the transport of prosaposin to the lysosomes. *J. Lipid. Res*, 1999. 40(9): p. 1593-603.

[125] Luberto, C. and Y.A. Hannun, Sphingomyelin synthase, a potential regulator of intracellular levels of ceramide and diacylglycerol during SV40 transformation. Does sphingomyelin synthase account for the putative phosphatidylcholine-specific phospholipase C? *J. Biol. Chem*, 1998. 273(23): p. 14550-9.

[126] Kornfeld, S. and I. Mellman, The biogenesis of lysosomes. *Annu. Rev. Cell Biol*, 1989. 5: p. 483-525.

[127] Igdoura, S.A., et al., Nonciliated cells of the rat efferent ducts endocytose testicular sulfated glycoprotein-1 (SGP-1) and synthesize SGP-1 derived saposins. *Anat. Rec*, 1993. 235(3): p. 411-24.

[128] Misumi, Y., et al., Novel blockade by brefeldin A of intracellular transport of secretory proteins in cultured rat hepatocytes. *J. Biol. Chem*, 1986. 261(24): p. 11398-403.

[129] Lippincott-Schwartz, J., et al., Rapid redistribution of Golgi proteins into the ER in cells treated with brefeldin A: evidence for membrane cycling from Golgi to ER. *Cell*, 1989. 56(5): p. 801-13.

[130] Yuan, L. and C.R. Morales, A stretch of 17 amino acids in the prosaposin C terminus is critical for its binding to sortilin and targeting to lysosomes. *J. Histochem. Cytochem.* 58(3): p. 287-300.

[131] Garnier, J., D.J. Osguthorpe, and B. Robson, Analysis of the accuracy and implications of simple methods for predicting the secondary structure of globular proteins. *J. Mol. Biol*, 1978. 120(1): p. 97-120.

[132] Rost, B., G. Yachdav, and J. Liu, The PredictProtein server. *Nucleic. Acids Res*, 2004. 32(Web Server issue): p. W321-6.

[133] Helms, J.B. and C. Zurzolo, Lipids as targeting signals: lipid rafts and intracellular trafficking. *Traffic*, 2004. 5(4): p. 247-54.

[134] Ikonen, E., Roles of lipid rafts in membrane transport. *Curr. Opin. Cell Biol*, 2001. 13(4): p. 470-7.

[135] Simons, K. and D. Toomre, Lipid rafts and signal transduction. Nat Rev *Mol. Cell Biol*, 2000. 1(1): p. 31-9.

[136] Simons, K. and E. Ikonen, Functional rafts in cell membranes. *Nature*, 1997. 387(6633): p. 569-72.

[137] Brown, D.A. and E. London, Structure and origin of ordered lipid domains in biological membranes. *J. Membr. Biol*, 1998. 164(2): p. 103-14.

[138] Schroeder, R.J., et al., Cholesterol and sphingolipid enhance the Triton X-100 insolubility of glycosylphosphatidylinositol-anchored proteins by promoting the formation of detergent-insoluble ordered membrane domains. *J. Biol. Chem*, 1998. 273(2): p. 1150-7.

[139] Canuel, M., et al., Sortilin and prosaposin localize to detergent-resistant membrane microdomains. *Exp. Cell Res*, 2009. 315(2): p. 240-7.

[140] Lefrancois, S., et al., The lysosomal transport of prosaposin requires the conditional interaction of its highly conserved d domain with sphingomyelin. *J. Biol. Chem*, 2002. 277(19): p. 17188-99.

[141] Ni, X., M. Canuel, and C.R. Morales, The sorting and trafficking of lysosomal proteins. *Histol. Histopathol*, 2006. 21(8): p. 899-913.

[142] Shogomori, H. and A.H. Futerman, Cholesterol depletion by methyl-beta-cyclodextrin blocks cholera toxin transport from endosomes to the Golgi apparatus in hippocampal neurons. *J. Neurochem*, 2001. 78(5): p. 991-9.

[143] Austin, C., M. Boehm, and S.A. Tooze, Site-specific cross-linking reveals a differential direct interaction of class 1, 2, and 3 ADP-ribosylation factors with adaptor protein complexes 1 and 3. *Biochemistry*, 2002. 41(14): p. 4669-77.

[144] Jackson, C.L. and J.E. Casanova, Turning on ARF: the Sec7 family of guanine-nucleotide-exchange factors. *Trends Cell Biol*, 2000. 10(2): p. 60-7.

[145] Nie, Z., D.S. Hirsch, and P.A. Randazzo, Arf and its many interactors. *Curr. Opin. Cell Biol*, 2003. 15(4): p. 396-404.

[146] Shin, H.W., et al., BIG2, a guanine nucleotide exchange factor for ADP-ribosylation factors: its localization to recycling endosomes and implication in the endosome integrity. *Mol. Biol. Cell*, 2004. 15(12): p. 5283-94.

[147] Morales, C.R., Role of sialic acid in the endocytosis of prosaposin by the nonciliated cells of the rat efferent ducts. *Mol. Reprod. Dev*, 1998. 51(2): p. 156-66.

[148] Hiesberger, T., et al., Cellular uptake of saposin (SAP) precursor and lysosomal delivery by the low density lipoprotein receptor-related protein (LRP). *Embo J*, 1998. 17(16): p. 4617-25.

[149] Hermey, G., et al., Identification and characterization of SorCS, a third member of a novel receptor family. Biochem. *Biophys. Res. Commun*, 1999. 266(2): p. 347-51.

[150] Yamashita, K., et al., Characteristics of asparagine-linked sugar chains of sphingolipid activator protein 1 purified from normal human liver and GM1 gangliosidosis (type 1) liver. *Biochemistry*, 1990. 29(12): p. 3030-9.

[151] Ito, K., et al., Structural study of the oligosaccharide moieties of sphingolipid activator proteins, saposins A, C and D obtained from the spleen of a Gaucher patient. *Eur. J. Biochem*, 1993. 215(1): p. 171-9.

[152] Waring, A.J., et al., Porcine cerebroside sulfate activator (saposin B) secondary structure: CD, FTIR, and NMR studies. *Mol. Genet. Metab*, 1998. 63(1): p. 14-25.

[153] Fluharty, A.L., et al., Preparation of the cerebroside sulfate activator (CSAct or saposin B) from human urine. *Mol. Genet. Metab*, 1999. 68(3): p. 391-403.

[154] Tatti, M., et al., Structural and membrane-binding properties of saposin D. *Eur. J. Biochem*, 1999. 263(2): p. 486-94.

[155] Burkhardt, J.K., et al., Accumulation of sphingolipids in SAP-precursor (prosaposin)-deficient fibroblasts occurs as intralysosomal membrane structures and can be completely reversed by treatment with human SAP-precursor. *Eur. J. Cell Biol*, 1997. 73(1): p. 10-8.

[156] Laurent-Matha, V., et al., Procathepsin D interacts with prosaposin in cancer cells but its internalization is not mediated by LDL receptor-related protein. *Exp. Cell Res*, 2002. 277(2): p. 210-9.

[157] Meertens, L., C. Bertaux, and T. Dragic, Hepatitis C virus entry requires a critical postinternalization step and delivery to early endosomes via clathrin-coated vesicles. *J. Virol*, 2006. 80(23): p. 11571-8.

[158] Gruenberg, J. and F.G. van der Goot, Mechanisms of pathogen entry through the endosomal compartments. *Nat. Rev. Mol. Cell Biol*, 2006. 7(7): p. 495-504.

[159] Fratti, R.A., et al., Role of phosphatidylinositol 3-kinase and Rab5 effectors in phagosomal biogenesis and mycobacterial phagosome maturation arrest. *J. Cell Biol*, 2001. 154(3): p. 631-44.

[160] Koul, A., et al., Interplay between mycobacteria and host signalling pathways. *Nat. Rev. Microbiol*, 2004. 2(3): p. 189-202.

[161] Alves, L., et al., Mycobacterium leprae infection of human Schwann cells depends on selective host kinases and pathogen-modulated endocytic pathways. *FEMS Microbiol. Lett*, 2004. 238(2): p. 429-37.

Chapter 6

GOLGI APPARATUS FUNCTIONS IN MANGANESE HOMEOSTASIS AND DETOXIFICATION

Richard Ortega and Asunción Carmona
Cellular Chemical Imaging and Speciation Group, CNAB, CNRS, Université Bordeaux 1, CENBG, Chemin du solarium,
33175 Gradignan, France

ABSTRACT

Recent data suggest that the Golgi apparatus plays a key role in the homeostasis and detoxification of manganese. Manganese is an essential trace element but when high exposure conditions occur, manganese induces neurological symptoms in human. Manganese is also suspected to be an environmental risk factor in the aetiology of Parkinson's disease. However, the mechanisms regulating manganese homeostasis and detoxification in mammalian cells are largely unknown. Owing to the development of synchrotron radiation X-ray nano-chemical imaging, we revealed the specific accumulation of manganese in the Golgi apparatus of dopaminergic cells in culture.

At both physiological and subcytotoxic concentrations of manganese, we found that manganese was essentially located within the Golgi apparatus. At cytotoxic concentration of manganese, we found a large increase of manganese content in the cytoplasm and the nucleus of dopaminergic cells. Similarly, if the Golgi apparatus is altered using brefeldin A, manganese reaches the nucleus and cytoplasm in higher content. The accumulation of manganese in the Golgi apparatus could have a preventative effect because manganese could be removed by exocytosis. However, vesicular trafficking could be disturbed by high concentrations of manganese leading to neuronal cell death. We will discuss the mechanisms involving the role of Golgi apparatus alteration in neurological disorders triggered by manganese.

INTRODUCTION

Manganese is an essential trace element for living cells but is toxic when present in high concentrations. Chronic exposure to manganese is toxic to the brain, resulting in manganism

(Couper, 1837), a neurological disorder with similar symptoms to Parkinson's disease (PD). In addition, occupational exposures to manganese are suspected to increase the risk of PD (Gorell et al., 1999; Lucchini et al., 2007) as well as being associated with high iron in dietary intakes (Powers et al., 2003). The biochemistry of manganese containing bio-molecules is now quite well characterized however this is not yet the case for the knowledge of manganese intracellular distribution and trafficking (for review: Reddi et al., 2009). It is only in recent years that the Golgi apparatus (GA) has been discovered to be an important organelle in manganese homeostasis and detoxification (for review: Van Baelen et al., 2004).

The determination of trace element distributions at the subcellular level, such as in the GA, is a challenging task because it requires the use of analytical tools with high spatial resolution and very high sensitivity, to target and detect trace elements within the intracellular organelles. This is particularly well illustrated in the case of GA and the discovery of its function in trace metal homeostasis.

For instance, it was thanks to the development of SIMS, secondary ion mass spectrometry, and EELS, electron energy loss spectrometry, that calcium storage within the GA was discovered (Chandra, 1991; Pezzati, 1997). This pioneer work enabled further study of the role of this GA calcium store in cell functions (for review: Van Baelen, 2004). In a similar approach, owing to the recent development of nano-chemical imaging based upon synchrotron radiation X-ray fluorescence, we revealed the role of GA in Mn homeostasis and detoxification in animal cells (Carmona et al., 2010).

In addition to results of manganese imaging in animal cells at low and high Mn concentrations, we will present new data on GA alteration after Mn exposure, and discuss how GA fractionation due to manganese accumulation could be involved in neurodegenerative diseases such as Parkinson's disease.

METHODS

Cell Culture

Rat pheochromocytoma PC12 cells were used as *in vitro model* of dopaminergic cells (Greene et al., 1976). PC12 cells were routinely maintained in RPMI 1640 medium (2.0 g/L glucose, 2 mM glutamine) supplemented with 10% horse serum, 5% fetal bovine serum (Sigma), and 100 U/mL penicillin-streptomycin, at 37°C in a water-saturated atmosphere containing 5% CO_2.

For chemical nano-imaging, cells were cultured directly on sample holders specially adapted for X-ray fluorescence analysis as recently described (Carmona et al., 2008).

Briefly, about $2 \cdot 10^4$ PC12 cells were split directly onto a 2 µm thin polycarbonate foil treated with gelatin gel mounted on a ESRF-ID22 sample holder with a 5 mm hole. PC12 cells were treated with NGF (nerve growth factor) at 100 ng/mL every two days, during an 8 day period, in order to obtain neuronal-like differentiated chromaffin cells. Exposure to manganese was carried out after NGF treatment.

Treatment and Sample Preparation for Nano-Imaging

Cells were exposed to 100 and 300 µM of $MnCl_2$ over 24 hours, medium was replaced by fresh medium and cells rested over 24 hours in normal conditions.

In another instance, cells were cryofixed into liquid isopentane cooled by liquid nitrogen (-160°C) and freeze-dried at -35°C. This protocol allows preservation of the integrity of cellular morphology and chemical element distribution in the cells so sample preparation does not induce any morphological and/or chemical modification (Ortega et al., 1996; Carmona et al., 2010).

For the brefeldin A treatment, cells were exposed to 300 µM of $MnCl_2$ during 24 hours. After that, medium was removed and cells were incubated with fresh medium at 30 µg/ml of brefeldin A over a four hour period, as described by other authors also using PC12 cells (Greaves et al., 2010). Cells were then cryofixed and freeze dried as previously explained.

Synchrotron X-Ray Fluorescence Nano-Imaging

The experiments have been conducted at ESRF (European Synchrotron Radiation Facility) on the nano-imaging facility ID22NI. The experimental station is located at a distance of 64 m from the undulator source and at 37 m from the high power slits used as the secondary source in the horizontal direction. The synchrotron radiation is focused by an X-ray optical device consisting of two elliptically shaped mirrors acting in two orthogonal planes using the so-called Kirkpatrick–Baez geometry (Hignette et al., 2005).

The first mirror, coated with a graded multilayer, plays both the role of vertical focusing device and monochromator, resulting in a very high and unique X-ray flux (up to 10^{12} photons/s) at energies between 15 and 29 keV. In this case, the energy of the pink photon beam was 16.4 keV, the flux $3.4 \ 10^{11}$ photon/s and spatial resolution of 220 nm x 90 nm (VxH).

The sample, mounted in air on a piezo nano-positioner stage, is scanned through the focal plane while the spectrum of the emitted fluorescence is recorded with an energy dispersive detector. The latter consists of a collimated silicon drift diode detector (SII Nanotechnology 50 mm^2 Vortex) placed in the horizontal plane at 90° from the incident beam and 45° to the sample surface normal.

A dwell time of 1s or 0.5s was chosen per point and 5 pixels per micron, as a compromise between a good spatial resolution (200 nm) and reasonable acquisition times to preserve the sample from radiation damages.

Spectra for each pixel were recorded in list mode so data treatment could be performed off-line. The recorded spectra are fitted to obtain maps of the element content using PyMCA software (Solé et al., 2007). PyMCA allows element-imaging reconstruction from recorded original spectra and also allows for obtaining the spectrum corresponding to a selected zone of the original image.

By selecting zones corresponding to the Golgi apparatus, nucleus and cytoplasm of cells, the corresponding spectra can be obtained and the proportion of manganese in each compartment calculated.

Golgi Apparatus Labeling

PC12 cells were transduced with Organelle Lights reagent (Invitrogen). The reagent contains a baculovirus (BacMam technology) which allows the expression of autofluorescent proteins that are localized to specific subcellular compartments and organelles of mammalian cells.

In this experiment green fluorescent protein (GFP) was used to mark the Golgi apparatus (Organelle Lights O36215).

The targeting sequence is the Golgi-resident N-acetylgalactosaminyltra-nsferase-2. PC12 cells were transduced according to the manufacturer's instructions of Invitrogen.

One day after transduction, cells were either exposed, or not, to manganese at 300 µM and 1200 µM during 24 h, and brefeldin A at 30 µg/mL during 4 h. Two days after transduction, organelle fluorescence was observed in living cells using an epifluorescence microscope (BX51, Olympus, Tokyo, Japan) and an U-MNIB2 filter.

RESULTS

X-ray fluorescence images reveal that in unexposed PC12 cells (control) manganese is located in the perinuclear region, always on one side of nucleus, identified as the Golgi apparatus (Figure 1A). When cells are exposed to a subcytotoxic dose of manganese (100 µM), the same distribution is found (Figure 1B).

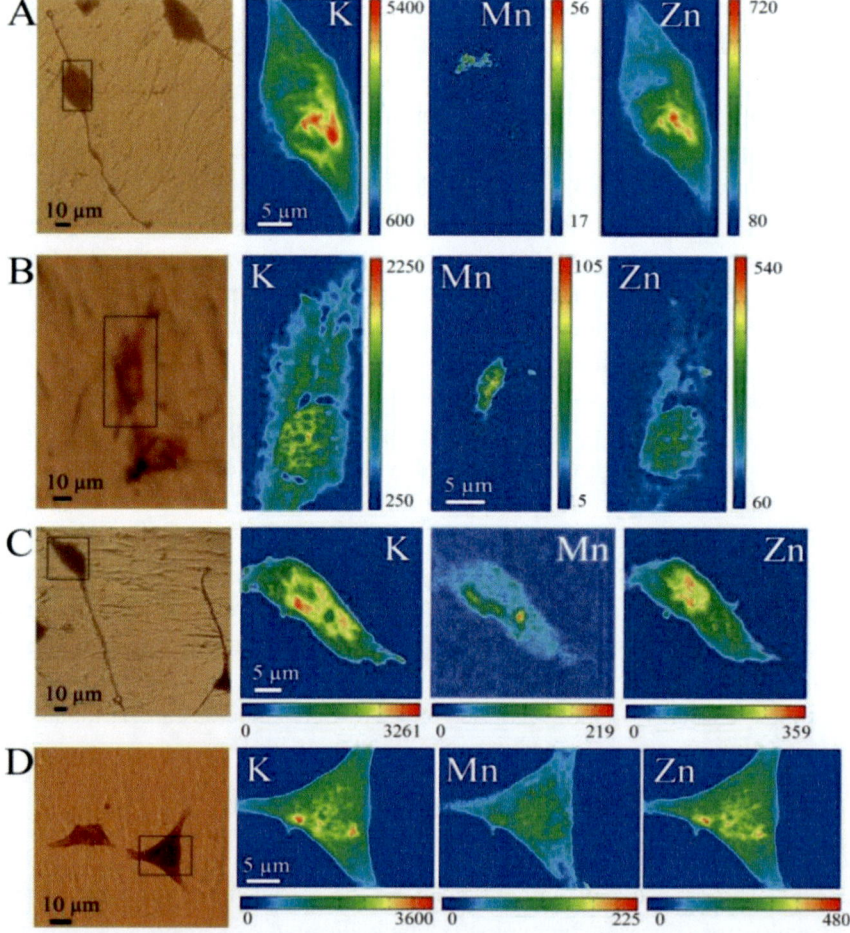

Figure 1. Potassium, manganese and zinc distributions in PC12 single cells. (A) Control cell. (B) Cell exposed to 100 µM MnCl2 during 24 h. (C) Cell exposed to 300 µM MnCl2 during 24 h. (D) Cell exposed to 300 µM MnCl2 during 24 h and 30 µg/mL of Brefeldin A during 4 h.

At higher manganese concentration (300 μM), Mn is also localized in the perinuclear region although the distribution is more diffuse than in figures 1A and 1B with Mn evidenced in the cytoplasm (Figure 1C), as also confirmed by quantitative analysis (Table 1).

The distribution of manganese is specific to the GA because other elements like K, or Zn are distributed differently into the cells. Potassium and zinc are quite homogeneously distributed, proportionate to the cell thickness, which explains the higher signal in the nucleus, as expected from the ubiquitous distribution of these elements in cells.

After alteration of protein transport to the Golgi apparatus by brefeldin A, manganese is redistributed within the cell showing a distribution similar to those of K and Zn (Figure 1D).

As explained in the 'Methods' section, using PyMCA software, we have calculated the proportion of manganese in GA, nucleus and cytoplasm (excluding GA) compared to the whole cell (Table 1).

For control cells, and cells exposed to 100 μM of Mn (non-toxic concentration), we found on average 60% of manganese in the GA, 20% in the nucleus, and 20% in the rest of cytoplasm.

For cells exposed to 300 μM of Mn (toxic concentration) we obtained on average 40% of manganese in the GA, 20% in the nucleus and 40% in the cytoplasm. When cells are exposed to brefeldin A we found that 45% of Mn accumulates in the nucleus and that 55% are in the cytoplasm.

Table 1. Absolute Mn content in cellular compartments of PC12 cells exposed to manganese at different concentrations and with brefeldin A.

	Manganese content (10^{-15} g)			
	Whole cell	Golgi	Cytoplasm	Nucleus
Control	< 1.2	< 0.72	< 0.24	< 0.24
100 μM MnCl$_2$	4	2.4	0.8	0.8
	Manganese content (10^{-15} g)			
	Whole cell	Golgi	Cytoplasm	Nucleus
300 μM MnCl$_2$	30	12	12	6
300 μM MnCl$_2$ + 30 μg/mL brefeldin A	80	-	44	36

In order to check for GA damage due to Mn exposures, PC12 cells were labeled with a GFP tagged to the GA (Figure 2). In control cells, not exposed to Mn, the GA is always located in the perinuclear region (Figure 2A).

After Mn exposure, evidence of GA fragmentation appears. This is particularly visible at high Mn concentration (Figure 2C), corresponding to 75% of PC12 cell death, but it is also observed in some cells at lower Mn concentrations (Figure 2B), corresponding to 25% PC12 cell death. GA fragmentation due to Mn exposure differs from GA disruption following brefeldin A exposure (Figure 2D).

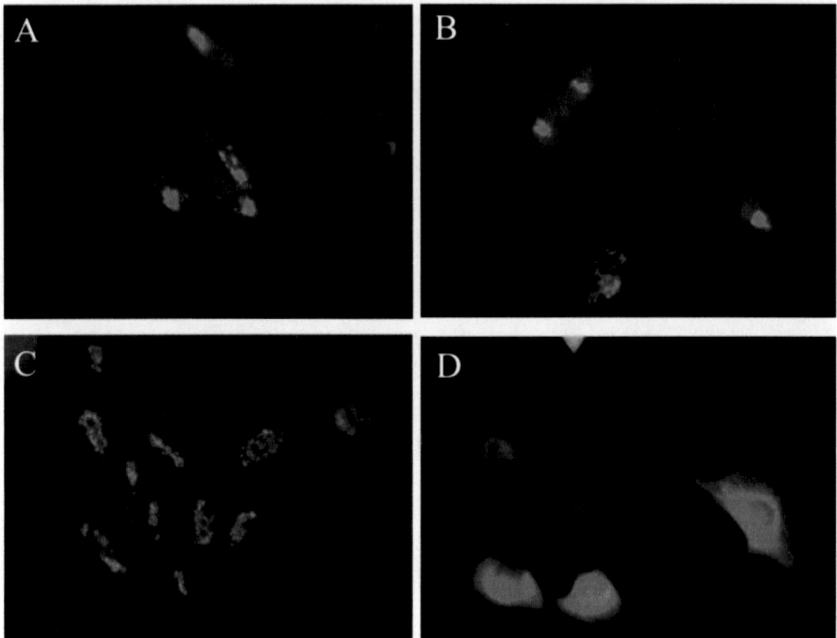

Figure 2. Golgi apparatus labeling with GFP Organelle Lights reagent in PC12 cells. (A) Control cells. (B) Cells exposed to 300 μM MnCl$_2$ during 24 h corresponding to the IC25. (C) Cells exposed to 1200 μM during 24 h corresponding to the IC75. (D) Cells exposed to 30 μg/mL of brefeldin A during 4 h.

DISCUSSION

Manganese is an essential trace element for human and chronic exposure to excess levels of Mn leads to neurotoxic symptoms; however, its subcellular distribution at physiological and toxicological concentrations remain to be elucidated. Using the direct chemical imaging nano-SXRF method, we provide evidence that Mn is located principally within the Golgi apparatus of PC12 cells cultured in normal conditions, without adding Mn to the culture medium (Figure 1A). In addition, with higher concentrations of Mn in the culture medium, 100 and 300 μM respectively, Mn accumulates in the GA (Figure 1B and 1C). Additionally, when Golgi trafficking is perturbed by brefeldin A, manganese is redistributed within the whole cell (Figure 1D).

Manganese Accumulation in GA

It is now well known that manganese is an essential element in GA apparatus functions. For example, manganese is required for the activation of galactosyltransferase, a GA enzyme, important in the processing of secreted proteins in many types of secretory cells (Witsell et al., 1990; Kuhn et al., 1991). In human cells, Mn can enter the cells by Dmt1, a Nramp protein that has a role in dietary iron and manganese uptake. Once into the cell, Mn can be directed to the GA through the SPCA1 and SPCA2 transporters, for secretory-pathway Ca^{2+}/Mn^{2+} transport ATPase, which are mainly targeted to the GA. These pumps supply the GA and other more distal compartments of the secretory pathway with the Ca and Mn necessary for the production and processing of secretory proteins. SPCA1 functions in both calcium and manganese transport (for review: Vangheluwe et al, 2009). SPCA1 is found in a compact perinuclear distribution, which corresponds to the location of the Golgi apparatus (Wootton et al., 2004). SPCA2 does not function in calcium transport but has specifically evolved for manganese homeostasis (Xiang et al., 2005). Immuno-cytochemical localization in human colon sections presents a typical apical juxtanuclear Golgi-like staining of SPAC2 (Vanoevelen et al., 2005).

Role of GA in Mn Detoxification

The first evidence that GA could be involved in Mn detoxification came from yeast studies. Pmr1, a P-type Ca^{2+}- and Mn^{2+}-transporting ATPase, is the yeast homolog for SPCA1 (Durr et al, 1998). Pmr1 transports Mn with high affinity and is specific in the Golgi stacks, the trans-Golgi network, and the secretory vesicles. It has been, suggested that in yeast, the excess manganese pumped into the Golgi by Pmr1, proceeds to exit the cell via secretory pathway vesicles that merge with the cell surface and release the manganese contents back into the extracellular environment (for review: Reddi, 2009). Also, in yeast, the Ccc1 polypeptide localizes to a Golgi-like organelle and can bind Mn in this organelle reducing the intracellular availability of the metal (Lapinski and Culotta, 1996).

Over-expression of Ccc1 in yeast cells resulted in reduced Mn cytotoxicity without lowering total accumulation of the metal. Although less is known about the role of SPCAs in manganese detoxification in animal cells, it has been hypothesized that, as in yeast, SPCAs can pump excess cytosolic manganese into the GA for its removal via the secretory pathway (Mandal, 2000; Wuytak, 2003).

Therefore, it can be expected that manganese in the GA would bind to a homolog of yeast Ccc1, which remains to be identified in animal cells. A similar mechanism of detoxification is also observed in plants such as *Arabidopsis thaliana*, a Golgi-based manganese accumulation resulting in manganese tolerance through vesicular trafficking and exocytosis (Peiter, 2007).

Our results support such a mechanism of detoxification, as manganese toxicity begins at 300 µM when Mn leaks out from the GA and reaches the cytoplasm and nucleus in a presumably toxic amount (Figure 1 and Table1). This is exemplified in the case of manganese co-exposure with brefeldin A, the amount of intracellular Mn being much higher than at 300 µM Mn alone (Table 1). This result strongly indicates an active role of GA and vesicular trafficking in Mn efflux and detoxification.

Manganese Alteration of GA and Neurodegenerative Diseases

There is increasing evidence that the alteration of the Golgi apparatus by fragmentation, and aggregation of misfolded or aberrant proteins, could contribute to the pathogeneses of neurodegeneration (Fan et al., 2008). The central nervous system is particularly vulnerable to manganese toxicity. High cytosolic concentrations of manganese are cytotoxic because they can interfere with magnesium-binding sites on proteins, and compromise the fidelity of DNA polymerases (Beckman et al., 1985).

On the other hand, incubation of mammalian cells, in the presence of millimolar concentrations of manganese, leads to a disruption of the normal membrane traffic, along the secretory pathway, most likely by interference with the motor proteins linking the membranes to the cytoskeleton (Towler et al., 2000). Our results evidence a fragmentation of GA at high Mn concentration (1200 µM) corresponding to the concentration of 75% of PC12 cells death (Figure 2C).

At lower Mn concentration the fragmentation of GA is less obvious (Figure 2B) when compared to control cells (Figure 2A). Manganese alters the GA morphology in a more distinctive way than brefeldin A (Figure 2D).

The interaction of Mn with the GA and secretory pathway is especially interesting regarding the possible involvement of Mn in the alteration of dopamine metabolism in the etiology of PD. There has been evidence that parkin, a protein mutated in the familial cases of PD, protects against Mn toxicity in dopaminergic cells but not in non-dopaminergic cells (Higashi, 2004).

Treatment with manganese resulted in accumulation of parkin protein in SH-SY5Y dopaminergic cells and its redistribution to the perinuclear region, especially aggregated Golgi complex, while in nondopaminergic Neuro-2a cells neither expression nor redistribution of parkin was noted. Our data could explain that the parkin protein is redistributed to the Golgi apparatus in dopaminergic cells in order to protect cells from Mn toxicity.

CONCLUSION

By direct observation of manganese distribution in dopaminergic cells at physiological concentration, we have evidence that this element is preferentially located to the GA. This result confirms the essentiality of Mn in GA functions.

Our results also suggest that the GA plays a role in manganese detoxification by storage of excess manganese and presumably exocytosis via the secretory pathway.

At toxic concentration of manganese in dopamine producing cells, overwhelming the binding capacity of GA, at least 2 mechanisms of manganese toxicity could apply: 1) the redistribution of manganese from the GA store to the nucleus, the cytoplasm and presumably the mitochondria where it could inhibit several essential cellular functions; 2) the direct alteration of GA functions and vesicular trafficking, resulting in interference with the motor proteins linking the membranes to the cytoskeleton. Regarding what is found in yeast, we hypothesize the existence of a polypeptide within the GA of mammalian and human cells responsible for Mn sequestration in GA and working in coordination with Mn transporting proteins such as SPCA1 and SPCA2 to control Mn homeostasis (uptake, sequestration, and efflux).

ACKNOWLEDGMENTS

Authors acknowledge the ESRF for beam time allocation and to ESRF staff for technical support. We are especially grateful to Dr. Peter Cloetens and Dr. Sylvain Bohic from ESRF.

REFERENCES

Beckman, R.A., Mildvan A.S., Loeb L.A. (1985) On the fidelity of DNA replication: manganese mutagenesis in vitro, *Biochemistry* 24, 5810–5817.

Carmona, A., Devès, G., Roudeau, S., Cloetens, P., Bohic, S., Ortega, R. (2010) Manganese accumulates within Golgi apparatus in dopaminergic cells as revealed by synchrotron X-Ray fluorescence nano-imaging. ACS *Chemical Neurosciences* 1, 194-203.

Carmona, A., Devès, G., Ortega, R. (2008) Quantitative micro-analysis of metal ions in subcellular compartments of cultured dopaminergic cells by combination of three ion beam techniques. *Anal. Bioanal. Chem.* 390, 1585-94.

Chandra, S., Kable, E.P., Morrison, G.H., Webb, W.W. (1991) Calcium sequestration in the Golgi apparatus of cultured mammalian cells revealed by laser scanning confocal microscopy and ion microscopy, *J. Cell. Sci.* 100, 747–752.

Couper, J. (1837) Sur les effets du peroxide de manganèse. *Journal de Chimie Médicale, de Pharmacie et de Toxicologie* 3, 233-235.

Durr, G., Strayle, J., Plemper, R., Elbs, S., Klee, S.K., Catty, P., Wolf, D.H., Rudolph, H.K. (1998) The medial-Golgi ion pump Pmr1 supplies the yeast secretory pathway with Ca^{2+} and Mn^{2+} required for glycosylation, sorting, and endoplasmic reticulum-associated protein degradation. *Mol. Biol. Cell* 9, 1149–1162.

Fan, J., Hu, Z., Zeng, L., Lu, W., Tang, X., Zhang, J., and Li, T. (2008) Golgi apparatus and neurodegenerative diseases. *Int. J. Dev. Neurosci.* 26, 523-34.

Gavin, C.-E., Gunter, K.-K., and Gunter, T.-E. (1990) Manganese and calcium efflux kinetics in brain mitochondria. Relevance to manganese toxicity. *Biochem. J.* 266, 329-34.

Gorell, J.-M., Johnson, C.-C., Rybicki, B.-A., Peterson, E.-L., Kortsha, G.-X., Brown, G.-G., and Richardson, R.-J. (1999) Occupational exposure to manganese, copper, lead, iron, mercury and zinc and the risk of Parkinson's disease. *Neurotoxicology* 20, 239-247.

Greaves, J., Salaun, Ch., Fukata, Y., Fukata, M., Chamberlain, L.-H. (2008) Palmitoylation and membrane interactions of the neuroprotective chaperone cysteine-string protein. *J. Biol. Chem.* 283, 25014-25026.

Greene, L.-A., and Tischler, A.-S. (1976) Establishment of a noradrenergic clonal line of rat adrenal pheochromocytoma cells which respond to nerve growth factor. *Proc. Natl. Acad. Sci. USA* 73, 2424-2428.

Higashi, Y., Asanuma, M., Miyazaki, I., Hattori, N., Mizuno, Y., Ogawa, N. (2004) Parkin attenuates manganese-induced dopaminergic cell death. *J. Neurochem.* 89, 1490-1497.

Hignette, O., Cloetens, P., Rostaing, G., Bernard, P., Morawe, C. (2005) Efficient sub 100 nm focusing of hard x rays. *Rev. Sci. Instrum.* 76, 063709.

Kalia, K., Jiang, W., and Zheng, W. (2008) Manganese accumulates primarily in nuclei of cultured brain cells. *Neurotoxicology* 29, 466-470.

Kuhn, N.J., Ward, S., Leong W.S. (1991) Submicromolar manganese dependence of Golgi vesicular galactosyltrasferase (lactose synthetase). *Eur. J. Biochem.* 195, 243-250.

Lapinskas, P.J., Lin, S.J., Culotta, V.C. (1996) The role of the *Saccharomyces cerevisiae* CCC1 gene in the homeostasis of manganese ions. *Mol. Microbiol.* 21, 519-528.

Lucchini, R.-G., Albini, E., Benedetti, L., Borghesi, S., Coccaglio, R., Malara, E.-C., Parrinello, G., Garattini, S., Resola, S., and Alessio, L. (2007) High prevalence of Parkinsonian disorders associated to manganese exposure in the vicinities of ferroalloy industries. *Am. J. Ind. Med.* 50, 788-800.

Mandal, D., Woolf, T.B., Rao, R. (2000) Manganese selectivity of pmr1, the yeast secretory pathway ion pump, is defined by residue gln783 in transmembrane segment 6. Residue Asp778 is essential for cation transport. *J. Biol. Chem.* 275, 23933-23938.

Morello, M., Canini, A., Mattioli, P., Sorge, R.-P., Alimonti, A., Bocca, B., Forte, G., Martorana, A., Bernardi, G., and Sancesario, G. (2008) Sub-cellular localization of manganese in the basal ganglia of normal and manganese-treated rats An electron spectroscopy imaging and electron energy-loss spectroscopy study. *Neurotoxicology* 29, 60-72.

Ortega, R., Moretto, P., Fajac, A., Benard, J., Llabador, Y., and Simonoff, M. (1996) Quantitative mapping of platinum and essential trace metal in cisplatin resistant and sensitive human ovarian adenocarcinoma cells. *Cell Mol. Biol.* 42, 77-88.

Peiter, E., Montanini, B., Gobert, A., Pedas, P., Husted, S., Maathuis, F.J-., Blaudez, D., Chalot, M., Sanders, D. (2007) A secretory pathway-localized cation diffusion facilitator confers plant manganese tolerance. *Proc. Natl. Acad. Sci. USA* 104, 8532-8537.

Pezzati, R., Bossi, M., Podini, P., Meldolesi, J., Grohovaz, F. (1997) High resolution calcium mapping of the endoplasmic reticulum-Golgiexocytic membrane system. Electron energy loss imaging analysis of quick frozen-freeze dried PC12 cells. *Mol. Biol. Cell.* 8, 1501–1512.

Powers, K.-M., Smith-Weller, T., Franklin, G.-M., Longstreth, W.-T.-Jr., Swanson, P.-D., and Checkoway, H. (2003) Parkinson's disease risks associated with dietary iron, manganese, and other nutrient intakes. *Neurology* 60, 1761-1766.

Reddi, A.R., Jensen, L.T., Culotta, V.C. (2009) Manganese homeostasis in Saccharomyces cerevisiae. *Chem. Rev.* 109, 4722-4732.

Solé, V.-A., Papillon, E., Cotte, M., Walter, Ph,. And Susini, J. (2007) A multiplatform code for the analysis of energy-dispersive X-ray fluorescence spectra. *Spectrochim. Acta* B 62, 63-68.

Towler, M.C., Prescott, A.R., James, J., Lucocq, J.M., Ponnambalam, S. (2000) The manganese cation disrupts membrane dynamics along the secretory pathway. *Exp. Cell. Res.* 259, 167-179.

Van Baelen, K., Dode, L., Vanoevelen, J., Callewaert, G., De Smedt, H., Missiaen, L., Parys, JB., Raeymaekers, L., Wuytack, F. (2004) The Ca^{2+}/Mn^{2+} pumps in the Golgi apparatus. *Biochim. Biophys. Acta* 1742, 103-112.

Vangheluwe, P., Sepulveda, M-R., Missiaen, L., Raeymaekers, L., Wuytack, F., Vanoevelen, J. (2009) Intracellular Ca^{2+}- and Mn^{2+}-transport ATPases. *Chem. Rev.,* 109, 4733–4759.

Vanoevelen, J., Dode, L., Van Baelen, K., Fairclough, R.J., Missiaen, L., Raeymaekers, L., Wuytack, F. (2005) The secretory pathway Ca^{2+}/Mn^{2+}-ATPase 2 Is a Golgi-localized pump with high affinity for Ca^{2+} ions. *J. Biol. Chem.* 280, 22800-22808.

Witsell, D.L., Casey, C.E., Neville, M.C. (1990) Divalent cation activation of galactosyltransferase in native mammary Golgi vesicles. *J. Biol. Chem.* 265, 15731-15737.

Wootton, L. L., Argent, C. C., Wheatley, M., Michelangeli, F. (2004) The expression, activity and localisation of the secretory pathway Ca^{2+}-ATPase (SPCA1) in different mammalian tissues. *Biochim. Biophys. Acta* 1664, 189–197.

Wuytack, F., Raeymaekers, L., Missiaen, L. (2003) PMR1/SPCA Ca^{2+} pumps and the role of the Golgi apparatus as a Ca^{2+} store. *Pflugers Arch.* 446, 148-153.

Xiang, M., Mohamalawari, D., Rao, R. (2005) A novel isoform of the secretory pathway Ca^{2+},Mn^{2+}-ATPase, hSPCA2, has unusual properties and is expressed in the brain. *J. Biol. Chem.* 280, 11608–11614.

INDEX

A

acetylcholine, 78
acid, viii, ix, 11, 20, 21, 34, 40, 41, 43, 44, 45, 54, 59, 61, 65, 66, 72, 75, 78, 79, 84, 86, 88, 113, 118, 119, 120, 121, 122, 125, 133, 134, 140, 143, 144, 145, 149
acidic, 23, 44, 49, 54, 120, 121, 125, 134, 143, 145
activation complex, 18
active site, 17, 22, 62, 120
acylation, 51, 84
adenocarcinoma, 112, 160
adenosine, 6
adenovirus, 102
adipocyte, 75
ADP, 34, 44, 46, 50, 71, 73, 75, 77, 78, 80, 82, 96, 101, 145, 146, 148
aetiology, x, 151
aggregation, 125, 126, 147, 157
albumin, ix, 27, 117, 126, 127, 135, 140
Aldrich syndrome, 77
algae, 27
alters, 41, 73, 158
amino, ix, 23, 48, 54, 59, 61, 65, 66, 69, 118, 122, 124, 125, 128, 133, 134, 140, 148
amino acid, ix, 48, 54, 59, 61, 65, 66, 69, 118, 122, 124, 125, 133, 134, 140, 148
amino acids, 48, 54, 65, 69, 122, 124, 133, 148
amphibians, 118
anchoring, 46
angiogenesis, 21
antibody, 28, 92, 125, 126, 127, 128, 129, 130, 131, 132, 135, 136, 138
antidepressant, 108
antigen, 41, 65, 83
antioxidant, 113

apoptosis, vii, 1, 2, 9, 10, 11, 16, 17, 18, 19, 20, 22, 23, 24, 25, 26, 37, 38, 39, 40, 41, 63, 75, 93, 97, 109, 110, 111, 112, 113, 114, 115, 116
apoptotic pathways, 16
AR, 31, 112, 114
Arabidopsis thaliana, 157
arrest, 6, 16, 21, 37, 149
arsenic, 115
arteries, 115
aspartate, 17
astrocytes, 40
atherosclerosis, 70
atmosphere, 152
atomic force, 69
atomic force microscope, 69
ATP, 13, 62, 73, 78, 80, 97, 98
attachment, 8, 44, 48, 97
autoantigens, 8, 29, 39
autoimmune diseases, 8
autosomal recessive, 70

B

basal ganglia, 160
base, 31, 63, 84, 146
bending, 48
bile, 61
biochemistry, 142, 152
biological activities, ix, 105, 108, 109
biological systems, 110
biosynthesis, 15, 16, 19, 20, 21, 26, 35, 36, 51, 62, 86, 87
birthmarks, 113
bladder cancer, 106, 109, 114
brain, 40, 116, 119, 141, 142, 151, 159, 160
breakdown, vii, 33, 94, 95
breathing, 109
budding, 3, 4, 6, 7, 8, 10, 12, 29, 32, 34, 45, 46, 47, 48, 55, 57, 59, 71, 72, 96, 102

C

Ca^{2+}, 17, 58, 83, 113, 125, 156, 157, 159, 160
calcium, 38, 54, 55, 58, 74, 77, 115, 126, 152, 157, 159, 160
cancer, 39, 106, 109, 113, 114, 115, 116, 149
cancer cells, 109, 113, 114, 115, 149
candidates, 95
carbohydrate, 119, 141
carbohydrates, 129
carcinoma, 109, 111, 112, 113, 114, 115, 116
cardiovascular disease, 71
cartoon, 5
casein, 13, 63, 65, 104
Caspase-8, 17, 110
caspases, 17, 19, 22, 23, 24, 25, 38, 110, 111
catabolism, 56, 121, 143
catalysis, 30, 121
catalytic activity, 58, 114
cation, 118, 141, 144, 146, 147, 159, 160
CD95, 41, 110
cDNA, 80, 126
cell cycle, 2, 5, 16, 21, 25, 26, 27, 28, 92, 93, 95, 97, 98, 100, 101, 102, 104, 112
cell death, viii, x, 2, 3, 16, 19, 21, 37, 38, 40, 41, 109, 110, 111, 112, 113, 114, 115, 116, 151, 155, 159
cell differentiation, 40
cell division, 21, 32, 92, 93, 98, 104
cell fate, 76
cell killing, 115
cell line, 4, 13, 28, 116
cell lines, 4, 116
cell membranes, 61, 148
cell movement, 78
cell signaling, 40, 60
cell surface, ix, 27, 30, 63, 75, 77, 117, 119, 122, 140, 157
central nervous system, 157
centrosome, 4, 5, 9, 25, 28, 41, 92, 98
cervical cancer, 106
challenges, 26, 71
chaperones, 81
chemical, x, 48, 59, 141, 151, 152, 153, 156
chemical properties, 48
chemicals, 107
chemiluminescence, 112
chemotherapy, 105
chicken, 141
China, 105, 107, 111
chitin, 34
CHO cells, 61
cholera, 8, 148
cholesterol, viii, 14, 15, 36, 43, 45, 48, 51, 56, 57, 58, 59, 60, 61, 63, 65, 66, 70, 71, 73, 77, 79, 80, 82, 83, 84, 86, 87, 119, 134
choline, 54, 57, 62, 84
chromatography, 87, 142
chromosome, 30, 80, 101, 120, 143
chromosome 10, 120
cilia, 41
cleavage, 17, 18, 19, 22, 23, 24, 25, 37, 38, 39, 60, 79, 110, 122
clinical application, 105, 106, 107
clinical presentation, 81
clinical trials, 106
cloning, 30, 35, 80, 141
clusters, 3, 19, 104
CO_2, 152
coagulopathy, 70
coding, 143
cognitive defects, 69
cognitive impairment, 70
collagen, 65
colon, 112, 113, 116, 157
colon cancer, 113, 116
compaction, 57
competition, 121, 128
complexity, vii, 1, 28, 118
complications, 106
composition, 45, 51, 55, 60, 69, 73, 87, 107, 119, 140, 141
condensation, 20, 51, 109
configuration, 60
congenital cataract, 69
connectivity, 8
consensus, 55, 58
conservation, 79
consumption, 24, 45
controversial, 8, 96
COOH, 126, 127, 135, 136
coronavirus, 27
correlation, 36, 85
crystal structure, 65
CS, 111, 112, 113, 114
culture, x, 129, 151, 156
culture medium, 129, 156
curcumin, 107, 113
cure, 106
cycling, 102, 148
cysteine, 17, 33, 55, 122, 126, 130, 145, 159
cytochrome, 16, 17, 23, 38, 109, 110
cytokinesis, 6, 18, 39, 101
cytometry, 109
cytoplasm, x, 3, 8, 10, 17, 18, 92, 97, 98, 109, 151, 153, 155, 157, 158

cytoplasmic tail, 49, 123, 138, 145
cytoskeleton, 4, 9, 10, 11, 12, 25, 31, 32, 33, 34, 69, 71, 80, 157, 158
cytotoxicity, 116, 157

D

database, 49, 62, 123
decoding, 125
defects, 16, 50, 69, 70, 76, 82, 143
deficiencies, 120
deficiency, 85, 120, 143, 144, 146
deformation, 46, 49, 50, 51, 71, 72, 76
degradation, 17, 22, 60, 63, 83, 85, 97, 119, 120, 124, 142, 143, 144, 159
dendrites, 141
Denmark, 101, 104
dephosphorylation, ix, 13, 52, 58, 65, 66, 77, 91, 98
depolarization, 115
depolymerization, 5, 11, 14, 25
derivatives, 12, 15, 20, 62, 65, 112
destruction, 113, 114
detection, 102
detergents, 60, 134
detoxification, x, 151, 152, 157, 158
diabetes, 108
diacylglycerol, viii, 11, 33, 43, 44, 45, 71, 72, 75, 81, 86, 147
dietary intake, 152
diffusion, 13, 71, 100, 110, 160
digestion, 129, 142
dimerization, 50, 64
diploid, 109
direct observation, 158
disclosure, 140
discs, 3
diseases, 8, 39, 87, 105, 108, 109, 152, 159
disorder, 69, 70, 88, 152
displacement, 68
dissociation, 22, 50, 65, 68, 73, 121
distribution, viii, ix, 13, 14, 19, 28, 29, 31, 45, 49, 54, 59, 83, 84, 91, 117, 132, 134, 135, 136, 143, 152, 153, 154, 155, 156, 157, 158
diversity, 27
DNA, 16, 18, 21, 38, 41, 92, 93, 97, 109, 115, 157, 158
DNA damage, 16, 18, 41, 97
DNA polymerase, 157
dopamine, 158
dopaminergic, x, 151, 152, 158, 159
Drosophila, 3, 7, 10, 11, 15, 16, 19, 25, 26, 27, 29, 37, 39, 71, 80, 84
drug delivery, 112
drug resistance, 105

dysplasia, 115

E

earthworms, 35
E-cadherin, 31
electron, 4, 27, 28, 106, 110, 125, 138, 152, 160
electron microscopy, 4, 27, 125, 138
elucidation, 140, 143
EM, 4, 37, 140
encoding, 29, 70, 73, 126
endothelial cells, 39, 70
enemies, 37
energy, 26, 49, 67, 106, 109, 114, 152, 153, 160
energy input, 49
engineering, 130, 140
environment, 45, 60, 121, 157
enzymatic activity, 12
enzyme, 12, 15, 18, 22, 24, 29, 38, 60, 63, 97, 103, 104, 120, 121, 141, 143, 144, 156
enzymes, ix, 2, 3, 12, 45, 52, 54, 55, 62, 63, 69, 70, 72, 86, 87, 98, 106, 117, 120, 140, 144, 145
EPC, 22
epithelial cells, 32, 48, 75, 138, 139
epithelium, 32
equilibrium, 32, 104
esophagus, 106
ester, 61, 70, 79, 113
ethanol, 109
etiology, 158
eukaryotic, vii, 1, 2, 3, 8, 9, 66, 68, 79, 92, 119
eukaryotic cell, vii, 1, 2, 9, 68, 119
evidence, ix, 3, 10, 12, 26, 28, 36, 66, 80, 96, 102, 105, 111, 121, 127, 136, 138, 140, 146, 148, 155, 156, 157, 158
evolution, 3, 92, 118
exaggeration, 2
exclusion, 59
exocytosis, x, 74, 151, 157, 158
exons, 142
experimental condition, 98, 126
export dynamics, 84
exposure, x, 121, 126, 151, 152, 155, 156, 157, 159
external environment, 45
extraction, 60, 70
extracts, 70, 100, 107
extrusion, 70

F

FAD, 23
families, 35, 81
family members, 37, 49, 118
fatty acids, 51, 60, 67

fibroblasts, 6, 69, 75, 79, 80, 83, 121, 124, 144, 145, 147, 149
fidelity, 157, 158
fission, viii, 12, 28, 33, 37, 43, 45, 50, 51, 54, 55, 72, 74, 80, 82, 88, 91, 92, 95, 100, 102
fluid, 139
fluorescence, 112, 127, 152, 153, 154, 159, 160
force, 69, 71
Ford, 49, 76
formation, viii, 7, 10, 11, 27, 29, 30, 32, 33, 39, 40, 41, 47, 48, 50, 51, 52, 55, 56, 60, 63, 64, 67, 69, 71, 72, 75, 76, 84, 87, 88, 91, 95, 96, 98, 102, 103, 110, 122, 128, 140, 142, 148
fragments, 9, 19, 24, 25, 95, 97, 98, 101, 103, 104
France, 106, 151
free radicals, 106
functional analysis, 71
functional changes, 115
fungal metabolite, 130
fusion, viii, 6, 8, 29, 43, 44, 45, 48, 50, 51, 55, 56, 62, 67, 68, 69, 70, 71, 73, 76, 78, 81, 91, 92, 96, 97, 98, 103, 127, 142

G

GDP, 46, 68, 73, 138
gel, 131, 152
gene expression, 19
gene regulation, 25
genes, 13, 19, 60, 70, 78, 118, 142
genetic disorders, 45
genetics, 143
genitourinary tract, 107
genome, 65
geometry, 153
Germany, 106, 133
glioblastoma, 116
glioma, 36, 109, 110, 114
glucose, 28, 31, 152
glucosidases, 20
GLUT4, 31
glutamine, 152
glycerol, 44, 51, 75, 120
glycoproteins, 2, 27, 120, 128, 130, 142
glycosaminoglycans, 2
glycosylation, ix, 2, 27, 62, 70, 81, 117, 123, 129, 130, 138, 140, 143, 159
grants, 99, 111
granules, 144
growth, 16, 19, 20, 31, 34, 37, 58, 62, 64, 72, 86, 93, 102, 104, 113, 114, 122, 141, 144, 152, 159
growth arrest, 37
growth factor, 72, 93, 102, 104, 114, 122, 141, 144, 152, 159

GTPases, 29, 46, 78, 85, 99
guanine, 44, 46, 68, 81, 85, 87, 123, 146, 148

H

halogen, 106, 107
haptoglobin, 27
healing, 30, 109, 114
heart disease, 108
hemoglobin, 114
hepatitis, 140
hepatocellular carcinoma, 112
hepatocytes, 27, 121, 145, 147
heterogeneity, 45
HIV, 140
HLA, 113
homeostasis, x, 14, 15, 19, 21, 37, 45, 54, 55, 66, 86, 151, 152, 157, 158, 159, 160
Hong Kong, 105, 111
host, 149
hub, 2
human, x, 9, 24, 29, 30, 35, 39, 40, 41, 45, 64, 65, 72, 78, 79, 80, 81, 82, 86, 102, 105, 109, 112, 113, 114, 115, 116, 142, 143, 146, 147, 149, 151, 156, 158, 160
human brain, 142
human genome, 65
human neutrophils, 72
hybrid, 128, 130
hydrogen, 23, 64, 65, 66, 106, 109, 115
hydrogen peroxide, 23, 106, 109, 115
hydrolysis, 11, 20, 33, 41, 46, 50, 52, 61, 63, 72, 97, 98, 120, 125, 138
hydrophilicity, 133
hydrophobicity, 133
hydroxyl, 57, 62, 106
hydroxyl groups, 57
hypothesis, ix, 103, 118, 126, 127, 133, 134, 135, 138

I

ideal, viii, 2, 19, 23, 26
identification, 14, 22, 24, 64, 125, 140, 142
identity, 15, 29, 58, 59, 68
illumination, 109
image, 2, 153
images, 4, 154
immunity, 24, 37, 39
immunofluorescence, 4, 125, 131
immunoglobulin, 26, 64
immunoprecipitation, 128, 130, 131, 134
in vitro, 7, 34, 50, 51, 56, 58, 62, 86, 103, 128, 133, 143, 152, 158

in vivo, 24, 103, 112, 114, 120, 138, 143
inducer, 22, 38, 110
induction, viii, 6, 24, 43, 76, 109, 111, 116
infection, 17, 18, 149
inflammation, 16, 21
inflammatory responses, 17
inheritance, 10, 28, 32, 97, 98, 100, 101, 103, 104
inhibition, 6, 8, 11, 22, 24, 38, 41, 51, 56, 66, 70, 94, 102, 110, 113, 124, 134, 138
inhibitor, 17, 38, 56, 68, 73, 94, 127
initiation, 37, 40, 46, 116
inositol, 12, 15, 33, 57, 59, 78, 81, 82, 83, 86, 88
insects, 63
insertion, 48, 50, 63
insulin, 4, 31, 60, 141, 144
integration, 26, 64
integrity, 12, 76, 148, 153
interface, 89
interference, 8, 22, 25, 94, 142, 157, 158
internalization, 122, 146, 149
interphase, 5, 7, 92, 98, 104
invaginate, 119
invertebrates, 118
ion transport, 69
ions, 126, 159, 160
iron, 152, 156, 159, 160
irradiation, 41, 112
ischemia, 116
isolation, 82, 93
isopentane, 153
isozymes, 78
issues, 7, 10, 19

J

Japan, 106, 154

K

kidney, 4, 28, 32, 33, 100
kill, 106, 109
kinase activity, 58, 77
kinetics, 25, 159
Krabbe disease, 120, 143

L

labeling, 126, 127, 128, 130, 131, 138, 139, 156
lactose, 159
LDL, 61, 86, 149
lead, vii, 1, 3, 16, 21, 22, 57, 64, 120, 133, 140, 159
leaks, 157
lecithin, 33
LED, 107, 113, 114
leucine, 58, 60
leukemia, 78
ligand, 36, 84, 110, 121, 128, 135, 137, 138
light, ix, 79, 105, 106, 107, 109, 111, 113, 114
light emitting diode, 107
lipases, 45
lipid metabolism, 45, 60, 64, 66, 85
lipids, vii, viii, 2, 6, 11, 12, 19, 20, 26, 34, 43, 45, 48, 51, 52, 53, 54, 56, 57, 63, 67, 70, 76, 81, 83, 86, 119, 120, 142, 143
lipoproteins, 60, 61
liposomes, 8, 48, 67
liver, 33, 35, 70, 74, 76, 143, 144, 149
localization, 12, 15, 18, 19, 20, 29, 31, 33, 34, 36, 38, 39, 50, 54, 57, 58, 63, 64, 66, 67, 74, 79, 80, 81, 82, 85, 87, 100, 101, 102, 110, 112, 119, 122, 135, 145, 148, 157, 160
low temperatures, 27, 134
LSD, 120
lumen, 14, 22, 36, 54, 61, 62, 119, 139
lung cancer, 106, 113
Luo, 112
lymphocytes, 144
lymphoid, 40
lysine, 58
lysis, 130
lysosome, 6, 20, 48, 49, 61, 77, 110, 145

M

machinery, viii, 3, 10, 12, 17, 22, 23, 35, 43, 45, 56, 77
macromolecules, vii
macrophages, 70, 80
magnesium, 20, 157
majority, 128
malignant tumors, ix, 105
mammalian cells, vii, x, 2, 3, 4, 7, 9, 10, 11, 13, 14, 18, 27, 28, 50, 51, 63, 76, 86, 92, 101, 119, 151, 153, 157, 159
mammalian tissues, 160
mammals, 15, 16, 20
management, ix, 105, 112
manganese, vii, x, 151, 152, 153, 154, 155, 156, 157, 158, 159, 160
mass, 59, 87, 152
mass spectrometry, 152
materials, viii, 43, 119
matrix, viii, 29, 91, 92, 100, 101, 103, 104
matter, iv, 39, 55
measurement, 112
measurements, 69
medical, ix, 105
medicine, 107, 111
MEK, 93, 94, 100

mellitus, 108
metabolic pathways, 60
metabolism, viii, 13, 15, 21, 36, 39, 41, 43, 45, 51, 57, 60, 61, 63, 64, 66, 73, 75, 79, 83, 84, 85, 88, 109, 114, 158
metabolites, 40
metabolized, 20
metal ion, 159
metal ions, 159
metaphase, viii, 91, 92, 95, 97
metastasis, 102
methylene blue, 107, 113
Mg^{2+}, 78
mice, 6, 16, 17, 24, 37, 83, 113, 115, 138, 139, 146
microinjection, 10, 95
microscope, 69, 110, 154
microscopy, 4, 27, 32, 57, 100, 104, 125, 131, 132, 138, 159
migration, 2, 10, 21, 25
mitochondria, 5, 7, 13, 16, 17, 23, 37, 51, 59, 109, 110, 111, 114, 115, 158, 159
mitochondrial damage, 109
mitochondrial DNA, 115
mitogen, viii, 58, 91, 100, 101
mitosis, viii, 5, 7, 10, 11, 18, 25, 26, 28, 39, 91, 92, 93, 94, 95, 96, 97, 98, 99, 100, 101, 102, 103, 104
mitotic index, 95
mixing, 67, 81
MMP, 109
models, 62, 64, 69, 97
modifications, 92, 97, 119
modules, 56, 65
molecular oxygen, 106
molecular structure, 140
molecular weight, 17, 145
molecules, 2, 11, 12, 14, 15, 16, 17, 20, 22, 24, 25, 26, 28, 50, 51, 80, 97, 106, 110, 111, 119, 125, 136, 140, 152
monoclonal antibody, 28
morphology, 4, 16, 22, 25, 32, 34, 41, 56, 57, 92, 153, 158
mosaic, 142
motif, ix, 13, 30, 35, 49, 50, 54, 55, 58, 62, 64, 65, 66, 68, 72, 74, 75, 78, 79, 81, 97, 98, 117, 123, 124, 140, 146
motor neuron disease, 80
mRNA, 24
mutagenesis, 76, 140, 158
mutant, 16, 37, 63, 68, 85, 97, 133, 134
mutation, 60, 67, 70, 121, 124, 133, 134, 143
mutational analysis, 126
mutations, 49, 61, 68, 69, 70, 133
mycobacteria, 149

myosin, 10, 31, 32

N

NAD, 82
nasopharyngeal carcinoma, 109, 111, 112, 114
NCS, 58
necrosis, 18, 20, 39, 40, 109
nematode, 64
neovascularization, 114
nerve, 122, 152, 159
nerve growth factor, 122, 152, 159
nervous system, 125, 157
neuroblastoma, 83
neurodegeneration, 115, 157
neurodegenerative diseases, 152, 159
neuronal cells, 121
neurons, 28, 141, 148
neutral, 20, 23, 37, 40, 41, 126
neutrophils, 72
Niemann-Pick disease, 61
nitrogen, 153
NMR, 63, 64, 76, 149
nucleation, 9, 11, 31, 77
nuclei, 33, 132, 159
nucleic acid, 106, 119
nucleus, x, 13, 16, 18, 19, 25, 38, 60, 151, 153, 154, 155, 157, 158

O

OH, 15, 34, 61, 65, 66, 76
oil, 152
oleic acid, 54
oligomerization, 7, 17, 95, 96
oligomers, 7, 95, 99
oligosaccharide, 27, 70, 120, 140, 149
organ, 70
organelles, viii, 5, 7, 13, 14, 16, 17, 23, 25, 31, 32, 43, 45, 55, 64, 65, 79, 91, 92, 110, 111, 118, 119, 152, 153
organize, 11, 123
organs, 144
overlap, 14, 18
oxidative reaction, 106
oxidative stress, 15, 16, 23, 24, 37, 41, 71, 83, 115
oxygen, ix, 23, 105, 106, 107, 109, 110, 111, 112, 114

P

p53, 17
pancreas, 26
parallel, 15, 69, 110
pathogenesis, 84

Index

pathogens, 140
pathophysiology, 85
pathways, vii, viii, 15, 16, 17, 18, 20, 21, 23, 24, 26, 36, 37, 40, 44, 46, 49, 52, 54, 59, 60, 70, 72, 93, 96, 99, 103, 109, 110, 112, 115, 116, 118, 119, 149
penicillin, 152
peptide, 72, 87, 147
peptides, 129
permeability, 37, 109
permission, iv, 94, 126, 127, 129, 131, 132, 135
permit, x, 118, 119
peroxide, 23, 109, 115, 159
pH, 20, 75, 110, 119, 125, 126, 130
pheochromocytoma, 152, 159
phosphate, viii, ix, 12, 13, 19, 20, 21, 34, 39, 40, 43, 44, 49, 51, 52, 56, 72, 76, 78, 84, 85, 87, 117, 118, 119, 121, 135, 141, 142, 144, 145, 146, 147
phosphates, viii, 43, 45
phosphatidylcholine, 11, 20, 33, 44, 71, 73, 79, 86, 147
phosphatidylethanolamine, 51
phosphatidylserine, 51, 85, 109
phosphoinositides, 12, 34, 74, 82, 84
phospholipids, viii, 2, 10, 43, 45, 52, 56, 57, 63, 70
phosphorylation, viii, 2, 5, 8, 12, 13, 15, 35, 36, 43, 55, 58, 63, 65, 67, 79, 80, 82, 85, 91, 92, 95, 96, 97, 101, 144
photons, 106, 153
photosensitivity, 107
physical interaction, 60
physical properties, 45
PI3K, 21
pituitary tumors, 114
plants, ix, 8, 92, 105, 109, 114, 118, 157
plasma membrane, ix, 2, 8, 9, 12, 14, 20, 21, 22, 23, 25, 27, 33, 39, 45, 72, 73, 77, 79, 80, 83, 84, 87, 117, 118, 119, 142, 145
plasma proteins, 144
plasminogen, 121, 144
platelets, 82
platform, viii, 2, 3, 9, 11, 19, 23, 26
platinum, 160
PM, 28, 31, 45, 46, 47, 49, 53, 55, 57, 58, 59, 60, 61, 63, 64, 66, 69, 70, 112, 113, 114
point mutation, 133
point of origin, 118
polar, 50, 51
polarity, 9, 25, 30
polarization, 10, 25
polycarbonate, 152
polymerization, 25, 49, 50, 85
polypeptide, 126, 157, 158

pools, viii, 43, 51
population, 10, 33, 35, 57, 80
porphyrins, 112
port-wine stain, 113
potassium, 128
preparation, iv, 153
preservation, 153
primary function, vii
priming, 67, 78
principles, 145
programming, 71
prokaryotes, 118
proliferation, 19, 20, 93
proline, 95, 132
promoter, 24, 41, 50, 65, 72
prophase, viii, 91, 95, 100
protein family, 37, 54, 71, 73, 80
protein kinase C, 20, 40, 55, 79
proteoglycans, 2
proteolysis, 73, 84, 125
pumps, 156, 160
purification, 141

Q

quantification, 138
quartz, 106, 107

R

Rab, 8, 68, 85
radiation, x, 111, 113, 151, 152, 153
Radiation, 153
radiation damage, 153
radiotherapy, 105
reactions, 21, 62, 106
reactive oxygen, ix, 23, 105, 106, 109, 110, 111, 112
receptors, ix, 17, 21, 23, 38, 39, 40, 48, 49, 61, 76, 103, 117, 118, 119, 121, 122, 123, 124, 128, 141, 142, 144, 145, 146
recognition, 34, 79, 80, 81, 87, 121, 140, 141, 146
reconstruction, 5, 6, 153
recruiting, viii, 12, 43, 122
recurrence, 105
recycling, 2, 11, 31, 84, 119, 124, 142, 148
redistribution, 60, 61, 102, 130, 148, 158
regeneration, 31
regrowth, 92, 97, 98, 101
relatives, 38
relevance, 71, 95
replication, 158
residues, ix, 2, 17, 49, 50, 54, 62, 63, 64, 65, 66, 69, 87, 118, 120, 121, 130, 132, 140, 141, 144
resistance, ix, 105, 117

resolution, 28, 64, 152, 153, 160
response, 2, 13, 16, 17, 18, 23, 24, 25, 37, 40, 44, 51, 59, 64, 66, 70, 75, 79, 83, 110, 115, 125
reticulum, vii, viii, ix, 1, 2, 27, 31, 34, 35, 37, 43, 73, 74, 78, 79, 80, 82, 83, 84, 85, 86, 88, 91, 92, 97, 100, 101, 104, 110, 117, 118, 159, 160
retinopathy, 114
ribosome, 22
rights, iv
risk, x, 151, 152, 159
risks, 160
RNA, 8, 22, 25, 94, 136, 142
RNAi, 48, 56, 58, 62, 63, 66
root, 28
routes, 13, 14, 20, 23
rowing, 66

S

SAP, 128, 147, 149
saponin, 125
saturation, 128, 137, 140
science, 113
second generation, 113
secretion, viii, ix, 25, 26, 30, 34, 43, 56, 58, 63, 117, 118, 128, 138, 145
seedlings, 113
segregation, 11, 74, 101, 136
selectivity, 159
self-organization, 29
serine, 13, 20, 41, 50, 61, 63, 65, 66, 78, 94, 95, 98
Sertoli cells, 28, 126, 139, 142, 147
serum, 58, 69, 152
shape, 10, 34, 45, 51, 54, 60
shock, 17
showing, ix, 65, 118, 137, 155
sialic acid, 120, 149
side chain, 120, 121, 140
signal transduction, 14, 19, 40, 72, 85, 148
signaling pathway, vii, viii, 16, 17, 18, 19, 20, 24, 25, 26, 36, 44, 60, 76, 93, 94, 99, 110
signalling, 39, 40, 83, 101, 122, 149
signals, vii, viii, 1, 16, 17, 18, 19, 22, 23, 24, 25, 38, 40, 48, 63, 80, 91, 92, 95, 98, 104, 146, 148
signs, 138
SII, 153
silicon, 113, 153
siRNA, 51, 124, 128
skeletal muscle, 119
skeleton, 32
skin, 106, 107, 143
SMS, 44, 75, 80
software, 130, 153, 155
solubility, 60

solution, 50, 63
somata, 141
somatic cell, 102
Spain, 91
species, ix, 3, 14, 23, 51, 105, 106, 109, 110, 111, 112
spectroscopy, 69, 133, 160
speculation, 12
sperm, 64
spindle, viii, 10, 32, 91, 92, 95, 97, 98, 103, 104
spleen, 149
squamous cell, 112, 115
squamous cell carcinoma, 112, 115
stability, 8, 10, 41, 70, 73
stabilization, 32, 45
stars, 5
starvation, 58
state, 18, 22, 54, 106, 109, 129, 136, 138
states, 69, 144
stem cells, 40
sterols, viii, 15, 36, 43, 63, 66, 67, 76
stomatitis, 8
storage, 15, 110, 120, 138, 140, 147, 152, 158
stress, vii, 1, 3, 9, 10, 14, 15, 16, 17, 18, 19, 22, 23, 24, 25, 26, 37, 38, 41, 71, 81, 83, 115
stroke, 71
structural protein, 6, 7, 9, 19, 22, 23, 25
structure, iv, vii, viii, 1, 2, 3, 4, 6, 7, 8, 9, 10, 11, 12, 14, 18, 19, 20, 22, 24, 25, 26, 27, 28, 29, 30, 38, 40, 43, 45, 55, 57, 59, 64, 65, 66, 68, 69, 77, 79, 82, 84, 85, 86, 87, 88, 95, 100, 103, 104, 120, 122, 130, 132, 133, 134, 140, 143, 146, 147, 148, 149
substrate, 6, 18, 24, 44, 46, 50, 55, 64, 67, 80, 82, 120, 127, 143, 147
substrates, 18, 21, 55, 99, 120
sucrose, 134, 135
sulfate, 62, 120, 149
sulfonamide, 112
Sun, 34, 72, 73, 87, 112, 114
suppression, 56, 61, 62
surface area, 45
survival, viii, 2, 3, 16, 17, 19, 21
symptoms, x, 111, 151, 152, 156
syndrome, 30, 58, 69, 74, 75, 76, 78, 83, 85, 86
synthesis, vii, viii, 1, 2, 12, 13, 14, 15, 20, 21, 23, 33, 34, 35, 36, 41, 43, 45, 50, 52, 54, 56, 57, 60, 61, 62, 63, 64, 65, 66, 67, 70, 71, 72, 74, 76, 77, 78, 79, 85, 86, 88, 110, 127, 128, 134, 147

T

T cell, 39

target, 6, 17, 46, 63, 67, 68, 73, 95, 97, 100, 101, 110, 111, 115, 119, 126, 138, 140, 146, 152
technical support, 158
techniques, 32, 57, 104, 159
technology, 153
telophase, 6, 97, 98
therapeutic approaches, x, 118, 119
therapy, vii, ix, 105, 106, 109, 111, 112, 113, 114, 115
thoughts, 71
threonine, 13, 41, 50, 63, 81
time allocation, 158
tissue, 4
TLR, 63
TNF, 18, 22, 23, 24, 39, 41, 63
topology, 35, 76, 78, 103, 142, 143
toxicity, 109, 157, 158, 159
toxin, 8, 148
transcription, 24, 51, 60, 73, 78, 82, 97
transcription factors, 51, 60
transduction, 14, 19, 40, 72, 85, 148, 154
transfection, 126
transformation, 102, 147
translocation, 15, 35, 51, 58, 60, 66, 77
treatment, 15, 18, 22, 23, 24, 30, 54, 94, 106, 110, 114, 127, 128, 129, 130, 137, 140, 149, 152, 153
triggers, 50, 63
tryptophan, 132
tuberculosis, 140
tumor, ix, 18, 39, 62, 105, 106, 108, 109, 110, 111, 114, 115
tumor cells, 106, 109, 110, 111, 115
tumor growth, 62
tumor necrosis factor, 18, 39
tumorigenesis, 102
tumors, ix, 105, 106, 109, 111, 114
tungsten, 106, 107
turnover, 12, 40, 121

tyrosine, 54, 146

U

ultrasound, 111
uniform, 29, 57
urine, 149
urokinase, 144
USA, 1, 106, 159, 160
UVB irradiation, 41

V

vacuole, 34, 67, 77, 78, 81, 145
variations, 129
vector, 126, 135, 136
vesicle, 6, 8, 13, 29, 30, 31, 33, 35, 44, 46, 47, 50, 51, 52, 54, 57, 58, 59, 64, 67, 68, 69, 70, 71, 72, 73, 80, 81, 83, 85, 88, 95, 96, 102, 103, 104, 142
visualization, 29, 32, 103

W

water, 108, 120, 152
wavelengths, 107
Western blot, 127, 131, 135
wild type, 97, 126, 127, 131, 133, 139
Wiskott-Aldrich syndrome, 77
wound healing, 30, 109, 114

Y

yeast, 3, 7, 9, 10, 11, 12, 15, 34, 54, 64, 66, 67, 78, 118, 141, 145, 146, 157, 158, 159
yield, 57, 140

Z

zinc, 20, 55, 74, 154, 155, 159